THE
ORIGINS
of Man and the Universe

THIS BOOK is the account of a spiritual enquiry into evolution, civilisation, our place in the universe and the structure of reality itself. It is a cosmology which relates to present science but takes us beyond the Big Bang. Charting the evolutionary work of consciousness on earth, it takes us back through man's psyche to our original state in eternity or God's mind. This is an extraordinary work that brings self-discovery to the exploration of the universe — an enlightening fusion of knowledge in which every reader can share, since the story of existence is the story of us all.

BARRY LONG was born in Australia in 1926 and is now recognised as one of the leading spiritual teachers in the West. For several years following his transcendental realisation he was engaged in a gnostic enquiry into life on earth and the nature of consciousness. This book is the summation of his many remarkable insights and realisations about life, death and our being in existence — a unique work of immense vision.

'A profound and spellbinding book by one of the world's great spiritual masters. This book is destined to become a classic.' THE PLANET

'Will probably stand for decades to come as an influence on the way we look at life.' NEW AGE GUARDIAN

THE
ORIGINS
of Man and the Universe

The myth that came to life

BARRY LONG

26 September 2016

BARRY LONG BOOKS

This Revised Edition first published 1998 by Barry Long Books
Reprinted 2001
First Edition published 1984 by Routledge & Kegan Paul

Barry Long Books
BCM Box 876, London WC1N 3XX, England
Box 5277, Gold Coast MC, Qld 4217, Australia
6230 Wilshire Boulevard, Suite 251, Los Angeles, CA 90048, USA
www.barrylong.org   email: info@barrylong.org

Cataloguing-in-Publication Data:
A catalogue record for this book is available from The British Library.
Library of Congress Catalog Card Number: 97-93601

ISBN 1 899324 12 7   [paperback]
ISBN 0 9508050 6 8   [hardcover]

Editor of first edition: Peter Kingsley
Editors of second edition: Clive Tempest and Jade Bell
Illustrations: Jorma Sauvala
Cover design: Rene Graphics, Brisbane
Star over lake photo: Linton Lowe
Photo of Barry Long: Andrew Hawdon
Printed in England on acid-free paper by Biddles Ltd

TO THE TRUTH

CONTENTS

FOREWORD

THIS BOOK IS the product of an intense period of introspection by
a man whom many regard as one of the spiritual masters of our
times. The story begins in 1978. Barry Long had been consciously
engaged in self-realisation for more than twenty years and now,
at the age of 52, he was about to explore the immense
knowledge unlocked by his realisations. As he began to write he
had no idea where it would lead. A flood of insights into human
nature simply took him from one subject to another. In fact he
had begun a project that would occupy him for the next five
years and encompass the entire story of existence. As he wrote
he went deeper and deeper into an exploration of the very
source of our consciousness, ultimately leading him to describe
the fundamental structures of reality itself.

The author's only source was his own spiritual realisation
and revelation. He refrained from consulting other authorities,
scientific, philosophical or spiritual. Through his own gnosis, or
direct knowledge of universal truth, he found he was accounting
for the host of perennial questions that have teased philosophers

since ancient times and still bemuse the leading physicists of our day. Not only that, but he was gathering all these insights together into one grand design, a mythic account of the work of consciousness on earth.

The many strands and themes of these writings were woven together to form a book and it was ready by 1983. The manuscript was shown to a leading publishing company in London, where Barry Long was then living. A little dubious about the lack of academic credentials but recognising the book's unique quality, the publisher took it on and printed five thousand copies. It was some years before that edition was out of print and many readers said it was a book that changed their lives.

For the present edition, which appears fourteen years after the first, Barry Long and his editors have given the text a thorough revision. The language and style is generally simpler and in many sections the exposition has been clarified. The structure of the book is very much changed, with a new approach to the division of the material into chapters and sections. An appendix has been added which provides further background to the contents. Some passages in the first edition have been deleted, others have been totally rewritten and there are many paragraphs that are completely new. The author has nevertheless remained true to the original revelation of the book, adding very little from his subsequent realisations yet bringing to it the greater refinement with which he naturally sees things now.

Clive Tempest

INTRODUCTION

*You are going on a mythic journey,
deep into the psyche, back before
the beginning of time*

INTRODUCTION

FROM THE BEGINNING of western culture the story of Genesis has been the fundamental myth of the West. It records that in the beginning there was a spiritual principle, an energy called Man, the one and only, and out of that principle came what we now call man and woman. As stated in the myth, woman was taken out of the one and only Man's side. Or as I would put it: woman is man's love and his love was taken from him and made separate. The one and only became the two of us and one thing led to another, right up to today and our problematic life in the modern world. Now our technological and scientific times are so distant from our origins that the scientists' Big Bang theory and the truth of the creation myth are treated as two separate and irreconcilable things. The trouble is that sooner or later all intelligent men and women want to know their origins and yearn for the unity of original myth. And eventually man realises that the purpose of his life is to take his love back into him, and for he and she in their union to return to the truth, the one and only.

As soon as love is taken from the original principle, the one and only disappears. But it is still there, in the consciousness of Man that all men and women share. The knowledge of our origins is always within reach, but it is no good looking for it through the rational, scientific mind. That is too conditioned and formal. Our original state is unconditioned, free and invisible; it has not taken form. This book is about how we came to leave the original principle; and its purpose is to direct your attention back to that reality.

We only became what we are in our sensory bodies and our brains through the devolution of the eternal spiritual principle into matter through aeons of time. First the psyche was formed and Man began to exist as a psychic principle. The psyche is invisible, but energetically potent, a potential behind the scenes. A psychic principle has no physical embodiment, so gradually Man prepared a body for the psychic presence of man and woman to enter. But if man was to have a body, first of all what we call the species had to evolve in physical form so that all the apparatus of sense-perception would be ready for him. Then the intelligent, psychic power of man had to find a way of breaking through into the evolving senses. Finally, when everything was ready, man entered the earth as his physical home. How it was managed and how a spiritual principle became intelligent flesh and blood is the incredible story told in the early chapters of this book.

Man's view of the universe is determined by his brain and his sense-perception. In the central part of the book we descend further into the psyche to get behind the senses. I look into the levels of mind that structure and organise man's place in creation and I reveal knowledge of how existence itself is formed. From there we go on into the universe to discover the cosmic scheme behind the forces of evolution. We actually look into the ultimate mystery of how nothing becomes something. We are drawn step by step back towards the original point of consciousness, before time began. Finally, at the end of the book, I am able to describe the fundamental structures that determine our reality, the original ideas that eventually manifested as our life on earth.

I describe the book as myth because that is the only way to read it. You will not be able to comprehend it if you read it with the rational, scientific mind. It is a journey into the psyche. In writing the book I undertook that journey, really a descent through man's past, and I made discoveries as I went along. As you read you will follow the same descent and make your own discoveries. In going deeper, there is an expanding vision of what is and always has been. But on the way down you will hit strata of harder material. From time to time you may find that you have to stop reading or go back over what you have already read. Do not be discouraged.

As the book tells the myth of Man, the first principle operating in all men and women, I write of 'man' and use the masculine personal pronoun throughout, without deferring in my use of language to awkward conventions of the he/she variety. 'Man' refers to humanity as a whole, both male and female.

As you read you may ask how you can know if what the book says is the truth. The broad answer is that the truth, or the knowledge of life and death and of such a profound subject as the origins of man and the universe, is within you and everyone now at this moment — deep in the psyche. Specifically, I have access to such knowledge because I am a spiritual master and teacher. A spiritual teacher is someone who has taken conscious responsibility for the spiritual life of a section of humanity. One's knowledge of reality is determined by the responsibility one has taken for life. Appropriate to that responsibility and task I am permitted to enter the higher levels of mind described in this book. I write and speak from there.

As the book is a journey into the human psyche — your own psyche — you already 'know' all that is in it. As it contains the truth, you cannot be misled; unless you try to bring to it your judgmental rational mind which does not know the truth. If you do that your mind will lead you into your memory, your opinions and beliefs, and confusion. You will not enter the psyche or find the truth. You hear the beauty or truth of birdsong with your psyche but name the kind of bird with your mind.

As you read the first section of the book and begin the journey down you will find a good deal that is true in your own experience, and this will be your affirmation that it is the truth. But as you go deeper into the book and into the more obscure regions of the psyche, you will be less able to draw on the truth of your own experience. You will have to trust more and more in the part of you that is most real and most true. Do not try to work out what you are reading but read on and listen for the ring of truth. The ring or sound of truth is the only means we have of recognising what is real. Through it we receive the enlightening energy of the unknown and by-pass the surface mind whose frantic need to know and name is the impediment to enlightenment. Listening for the ring of truth will lead you to the extraordinary experience of realising that you did in fact already 'know' what your mind would try to tell you nobody could know.

I repeat, all that is in this book is already within you. Don't be put off by the mind that wants to judge, reject, form an opinion, argue, complain that the subject is too deep or say it does not understand. Read on. Let the words wash through you and over you. Read it as you would pause to look at the sea or the stars. Don't try to understand. Don't compare. Don't compete. There is nothing to prove. It will all come together within you; if not immediately, in time.

One

EVOLUTION

*The purpose of evolution at
any time is to make life on
earth more conscious*

EVOLUTION

MAN'S SPECULATIONS ABOUT his origins fall into two well-known categories — science's evolution theory, and the world's religious and mythical accounts.

Evolution theory, according to Darwin and his followers, describes man's animal development from the first micro-organism. No explanation is offered for his incomparable creative and artistic powers. According to the theory he is just a better kind of ape. On the other hand, biblical and other religious traditions report some kind of divine or mystical creation in which man appeared physically fully formed and imaginatively inventive as we know him today. The two approaches are irreconcilable — until the underlying truth connecting them is realised. Then the religious and mythical accounts are seen to be attempts at describing the astonishing emotional/creative beginnings of man — the element missing from the scientific theory.

As the religious stories pre-dated scientific rationalism by many, many thousands of years they could not avoid presenting a strongly emotional view: man's evolution since that time has been broadly from an emotional to a rational being. What is surprising today is that in spite of more than a century of science since Darwin, religious authorities continue to ignore the clamour in man's mind for a rational and intelligent explanation of religious phenomena in the light of evolution theory. This omission, and the implicit inability of organised religion to understand the truth of its own doctrines, probably accounts more than anything else for the decline in the West of religion's credibility and influence. Modern rational man, who depends on material fact for his daily bread and his motor car, has outgrown the need for emotional explanations of the origins of things and will not accept that human bodies, bread, or cars are suddenly created without a physical cause. And why should he? In a rational world such propositions are patently absurd. Yet intelligent and earnest men continue to commit themselves to the religious creation idea. This can only mean a massive misunderstanding has occurred somewhere along the line.

One would imagine that the two sides, having set themselves up as authorities in such matters as the origins of man, would be concerned with trying to find out what that all-important error or misunderstanding between them was, or is, instead of pursuing their mutually exclusive dogmas to the confusion of everyone who just wants to know the truth. Religion argues from a position of faith. Science argues from a position of fact. With less emphasis on positions, faith can be factually described and fact faithfully sourced to original causes.

These two great opposing ideas — evolution theory and creation theory — actually represent the two sides of man, his mystifying dual nature: the sensory physical, and the vitally creative. Without doubt man is an animal; and yet he is uniquely creative. How, and in what circumstances, did an animal become independently creative, numerically inventive, emotionally resonant? And where is the evidence? Under this kind of scrutiny, it is fairly obvious that neither evolution theory nor scriptural theory is really complete without the other. And yet to believe in either of these perennial opposites seems automatically to deny the possibility of the other. As it is impossible to argue sensibly against evolution as regards man's animal development, evolution theory has become almost universally accepted as describing his whole ancestry — even though it accounts for only half of the man which every scientist, every human being, experiences himself to be. To date, due more to the absence of an acceptable rationale than to any great personal conviction about his descent/ascent from apes, modern rational man largely has had to lay aside the fascinating scriptural and mythical possibility (talk of immortality, higher powers, visions and the rest) like a fondly remembered but impossible childhood dream.

Rationality, science — intellectual materialism — have triumphed over man's deeper feelings and longings; and on the evidence adduced, rightly so. But the trouble with a theory — Darwinian theory, religious theory, any theory — is that it always remains a partly observed fact. That is, it is not yet a principle, not immediately demonstrable, like for instance the law of gravity. No matter how carefully and in what detail phenomena are studied from an exclusive point of view — such as the animal ancestry of man which excludes the vitally creative side, or scriptural theory which excludes the rational side — the results are always partial, incomplete. And man remains as he is today, perhaps factually persuaded but faithfully unconvinced.

Evolutionists and naturalists like Darwin and so many others in our own time have contributed enormously to our understanding and appreciation of nature as the perfect system of perpetual motion. But what of the evolutionist himself? — not his organism, but the intelligence, the creative brilliance or genius with which he cognises the nature-system. What would be the result if he observed and catalogued this unique system of intelligence in himself, with the same patience and dedication as he studies the formal, natural world around him?

By demarking a point some tens of thousands of years ago when modern creative man (Homo sapiens) emerged out of the animal lineage of flesh and instinct, science tacitly acknowledges the occurrence of a unique and extraordinary event in time. It affirms that somewhere, sometime during the countless ages of the development of Homo (the biological order of primates which includes monkeys and apes) man acquired his sapience (sense of knowing discernment) that unprecedented humanising addition to his animal nature; and that this sapient factor is the other half of him, the missing link, in the otherwise elegant scientific theory.

THE INTELLIGENCE BEHIND EVOLUTION

To understand evolution before man appeared on earth, before there were selves which we as man are today, it is necessary to grasp that Man was a principle, not yet an entity. Man was the principle of intelligence behind all evolving forms of life — the intelligence that organised the evolution of the physical species. It may help to think of him as being already within or 'behind' the brains of the dinosaurs who preceded him, deep within the vast unconscious that was life at that time.

Through the evolution of the physical species on earth, Man as the principle of intelligence was able to evolve the human body to his own specifications. He did this over an enormous tract of time by gradually entering a partly developed physical body prepared for him by the earlier evolution of the rest of the species. The preparation of this body was the fundamental task of evolution. Although a very primitive body, it was already functional. Advancing into its brain as a relentless, intelligent pressure, Man forced the body to evolve its human form, uniquely suited to the creative and selfconscious capabilities which we as his selves express today. This story of the evolution of Man and the species has never been told before and I will now relate it in detail.

Every form ever taken by the earth's evolving species, from micro-organisms to primates, was a branch of the tree of life along which travelled the impulse that finally appeared as modern creative man — the selfconscious or self-reflective culmination of all the species. The first simple life-forms that appeared on earth were like the spring shoots on an otherwise invisible though gigantic trunk within the primordial psyche. By the time these life-forms appeared, this trunk was a superbly advanced network providing an acutely sensitive internal system of communication between all living things.

Evolution was mostly an exteriorising process. Everything came largely from within and was managed from within. Evolution theory, being based on external observation, has no knowledge of any of this. It misses the essential point that evolution was a dual process. It was involuntarily physical and biological on the surface and intelligently self-creative and vitally persistent underneath and within. Both functions ultimately were to merge as the body and intelligence of Homo sapiens, inventive man.

We assume from what evolution theory tells us that the environment on the surface of the earth must have been extremely hostile for the first life forms. Unarguably, the difficulty was for life to obtain a foothold and to hold on to it. What we are not told is that behind the simple life-forms was a vast psychic network — a psychic brain controlled by its own incredible intelligence. However much the tormented earth writhed, and its elements lashed, the brain responded from within with a viable organic mutation that somewhere, somehow, adapted and survived. What vital life within lacked in physical capability and resource, the brain made up for in timeless persistence and a genius for adaptation.

The psychic brain provided the struggling life-forms with co-ordinated direction. Using every form of life on the surface as a sensor, it knew exactly the conditions operating and developing anywhere on earth, wherever life had managed to get a hold. The brain evaluated this information and through its psychic infrastructure relayed back to all relevant life-forms the necessary adaptive changes and new procedures to be started. Covering the entire region from within, its responses were immediate and localised over the entire globe. Thus was all life able to keep ahead by anticipating geophysical and climatic extremes. Today, this continues: nature's anticipation of seasonal extremes and geophysical violence such as impending earthquakes — so evident among plants, insects and other animals — and all the other wonderful examples of nature's genius, is due to the use of the same subliminal psychic network. Similarly, the balance of advantage between predator and prey, which is essential to any life and survival pattern, was and is maintained with consummate delicacy.

Living organisms did not (and still do not) evolve in the external materialistic way that science and evolution theory maintain. The external, mechanistic and chemical changes occurring in the cells, tissues and organs were merely the effects of this amazing psychic system of interchange of information between every living form. The stick insect, to mention just one of the milliards of living natural wonders, did not evolve by sitting on twigs and being gobbled up over and over again while waiting for its colour and shape to change automatically, as the scientific explanation suggests. The whole miracle of camouflage, interdependence and the evolutionary survival of all the species was and still is to a large extent the work of the psychic brain and the intelligence behind it. This came first, not the chicken or the egg.

THE WORLD OF THE PSYCHE

All life that dies on earth survives in the psyche.

Probably the most tragic loss or omission for mankind due to scientific and intellectual externalism is the knowledge that within the human psyche — within the brain of the reader now — exists the astonishing, living, inner world in which humanity and all the species survive death in vital energetic form. This, the much misunderstood psychic world, was not always there.

Today, reconstituted and refined, it is teeming with life. In the beginning, before life on earth, it was merely a field of vital potential — the pre-conscious; a field of electrical force which today still provides the electrical current for the physical brainwaves between parts of our brain.

Life when it first started on earth continued in the most elementary form for an inordinately long period before suddenly exploding into infinitely more complex and varied organisms. Fossil evidence, or any correct reconstruction of the growth of early life, will confirm this. The delay occurred while vital energy from life-forms dying on earth was building up in the psychic field. As life perished on the surface, vital energy returned via the trunk system and gathered in the vital field before finally recirculating back up to the surface. At first the surviving psychic energies were extremely weak and faint, but eventually they combined in the psychic field to form a great reservoir and pressure of vital life. What had been a mere seep of recirculating life-sustaining energy became an intense instinctive drive from within for re-existence or re-experience. Surface life then burgeoned.

In the beginning, life which had survived death on earth had no particular form in the psychic field. Life on the surface was too simple for the surviving psychic energies to have individual shape or conscious identity. But as the many developing species became more abundant and firmly entrenched on earth, vital replicas of the organisms began appearing in the psychic mass — psychic doubles or doppelgangers.

The psychic field or world as it was now becoming, contained the energy or psychic double of every experiment at life on earth. But only so long as a species managed to survive and continued to evolve on the surface did the psychic doubles have a permanent psychic structure within. Otherwise they tended to disintegrate and vanish into the mass. Compared to their physical counterparts, the psychic doubles possessed a more intense and enduring reality. They did not die (as we apprehend death) but constantly changed their energetic frequency or image. Half a life-form, such as half a healthy living dog, is impossible on earth

15

but in the energetic psychic world, an amazing world which exceeds the parameters of ordinary conceptual imagery, halves and pieces of developing life-forms are in the natural order of things, as are 'impossible' combinations of life-forms that had or have no real future.

THE EMERGENCE OF THE BRAIN

The purpose of evolution at any time is to make life on earth more conscious.

The first life-forms possessed the minimal consciousness — instinct. Instinct can be called the consciousness of the unconscious. It is the experience or knowledge of having survived life and death, plus the urge to live or experience again. All life is instinctive because all life has survived death.

For scores of millions of years instinct as the desire to live and experience again was a sufficient dynamic to keep the species evolving under the co-ordinating inner guidance of the psychic brain. But there was a limit to how far instinct could serve the purpose of increasing consciousness in the species. Instinct is concerned with survival and the blind repetition of experience. It is instantaneous, immediate in its response. It could never provide sufficient pause in the awareness of an organism to sustain reflective selfconsciousness — the next evolutionary requirement. For that, more sophisticated bodies or organisms had to be evolved.

This next stage of evolution involved an extraordinary psychic performance which led to the development of the mammalian body. The psychic brain began to exteriorise itself. Physical versions of the brain started appearing in some of the species. The effect was to transfer into the bodies of the organisms themselves many of the internal growth and survival functions previously controlled by the psychic brain. This exteriorising of the psychic brain into innumerable creature brains was made possible through the simultaneous handover from psychic to genetic control.

Genetics, the scientific study of hereditary variations in organisms, began at the beginning of the twentieth century. The genetic code, the handwriting of the psychic brain of a hundred million years ago, was discovered by science in the 1950s. Scientists have been busily trying to decipher the code and they know, to their continued astonishment, that these intelligent biological instructions or ideas are actually recorded in the DNA of the cells. The genetic code contains all the data an organism will ever need for its life, propagation and eventual deterioration or devolution, even anticipating hybrid variations.

The transfer from psychic to genetic control not only made possible the evolution of the highest class of vertebrates, the mammals, and therefore the primates, but also prepared the way for probably the most evolutionary climacteric of all. It would eventually allow man to begin the step up out of the psyche into the world, into his by now potentially prepared physical body. All he would have to do would be to gradually realise (enter consciously) the various senses in the exteriorised brain that had been partly developed for him by the rest of the species. In so doing he would be helping to evolve the senses and his physical body to the point of readiness for his final triumphal entry into the world as a selfconscious being.

THE BODY OF PRIMITIVE MAN

Primitive man's body, when the drive to selfconsciousness began, would have been barely recognisable beside the human body today. All his senses were then present in an external body, but only in rudimentary form. He had only just evolved to the physical means of sight. Sight, as we know it, was the last physical sense to develop.

It took several hundred thousand years to realise the senses; that is, for the intelligence within to exteriorise itself; or, put another way, for man to come to his senses. Meanwhile primitive man lived mostly within his vital psychic double: he did not realise he possessed a physical body. His body performed on the earth in a totally instinctive way while the vital, feeling part of him lived a completely interiorised (though parallel) psychic existence.

Primitive man was not an individual as we today can experience ourselves. We have separative feelings — feelings of separate bodies, separate experiences. Primitive man experienced the world as himself. He could not think. He did not smell or hear things through his nose or ears, or use his memory to identify these perceptions as we do. He used a completely subjective internal sense-relatedness which was truly remarkable compared to our selfconscious understanding.

Primitive man was the consciousness-centre of the vital or psychic brain, the most advanced and sensitive point in the world organism. He was linked to his environment by an interiorised system of psychic sense (not neural or sense-perceptive sense) which kept him constantly informed in relation to his needs. His range of experience was greater than that of any other living thing. All earth-life served to inform him through the incredible maze of his psychic brain.

This is reflected in the story of Genesis in which man is given dominion over every living thing. Man had this control and superiority by virtue of his central part in creation as the power of intelligence in and behind all life. Science may some day discover that there is a connection between this and why more than half of modern man's brain is relatively inactive, or used for very few of his known functions relative to its mass.

When primitive man needed food he knew psychically in which direction to go and hunt for it. This is the essence of instinct. The nearest food, whether plant or animal, communicated its presence to him through the psychic network of which he was the centre. Naturally, he did not always find or catch his prey; he might fall prey to another hunter, or the prey might have used

18

its natural instincts to camouflage or conceal its position when he got there. But the point is that primitive man knew, through being at the centre of a subjective world-data system, where to go to find what he needed without exterior sense-aids. It was the same when he needed a female. He knew where one was, ready to receive him. She might have been with another male or with a rival tribe; and if he persisted in trying to mate with her he might be killed. But he knew where to locate such a female (and she, him) through this extraordinary psychic system — and both took their chances. This original inner-sense system is present in our bodies today but has been corrupted by the modern mind's imagination.

Primitive man felt pain. But he did not know 'whose' pain it was. For us a headache is 'my' headache, my exclusive pain. But not for primitive man. When he felt pain he *was* pain. There was no sense of self to claim it as his. Pain was the same for all. When he fought with another man or animal he did not know which pain was his. Logically, it could be said it had to be his own pain he was feeling. But that was not his experience. He knew only that fighting was painful. He avoided it when possible. And he did not injure or destroy other things unless they stood between him and his needs or his survival. He fought instinctively to lead and to procreate because survival of the fittest was the only natural guarantee of the survival of all.

While modern man with his sense of self (his selfconsciousness) selects things to own apart from his pain — my child, my house, etc — primitive man was a corporate feeling of all the things he required to survive. To him it was not just his world. He *was* the world. He had a kind of broad tribal identity but not a communal existence in which he shared his female or anything else. She was neither his, nor anyone's. She was just female — a feeling, not an objective form; and to her he was the same. Neither she nor any other thing he needed ever appeared in his feeling world or consciousness until he had the feeling of desiring her or it. Between those feelings female had no existence for him, as male had no existence for her; she and he were just there, part of the scene, much the same as the pictures that hang on our walls or the many other objects that are just there in our lives until we want them or need to notice them.

Without a sense of self, primitive man was not identified with 'mine and ours'. His home was just for protection, nothing more. He was not identified with winning or losing. If he had to fight for what he needed he had no sense of victory or defeat. There were only the spoils of war to be used or the pain of his wounds to endure, until contentment, the absence of desiring anything, came again.

Primitive man's battles were actually fought within himself, within his feeling brain. While his outer body might have gone through the action of combat, he merely registered the sensations of what was going on in his internal world, which was as instantaneous in its communication to him as our sense-perceptions are today. The difference is that there were no images, no concepts, no forms. He had not realised his senses as his physical body, so his body could not exist for him, even though it might have existed for a hypothetical observer. Primitive man's battles were similar to our experience when we are at war with ourselves, feeling remorse, regrets, guilt, self-doubt, indecision, uncertainty, worry. Whereas we are frequently in two minds, experiencing all the pain or discomfort of conflict without any external contact, he was constantly in two feelings, or a multiplicity of feelings. These feelings were searingly intense and abundant in information compared to ours.

As we compare primitive man's state with ours, a striking difference is revealed. Primitive man did not want; he desired. These are not the same sensations. Desire arises from the need to survive and is fundamental to life. Wanting is discontent — a symptom of developing selfconsciousness. Wanting is violence, a necessary transition on the way to a sense of self.

Primitive man was not violent. He was savage and dangerous as a wolf or gorilla might be described. But we would not normally call these wild animals violent; certainly not in the sense of man's violence as reported daily by the media.

20

Primitive man's inability to want selfconsciously was part of evolution's amazing system of controls and checks which are permanently present to preserve the unswerving order of things. Man's evolution can never be interrupted by an external event which is not itself a part of the evolutionary scheme. Here was primitive man, an extremely dangerous physical and psychic force, lacking all self-control and actually at the centre or brain of things, made to behave ethically, almost responsibly, by the very simplicity of his existence. Being at the centre of the interior world in which he had the feel of all things in his environment, he neither injured nor destroyed needlessly. He fought, but no longer nor more destructively than necessary. He was not cruel, sadistic, competitive or exploitative. Paradoxically, the world was safe in his savage hands. Primitive man was the ideal environmentalist.

Primitive man, as the intelligent focus of all life on earth, was the receptor of the 'feelings' expressed by his environment. Today we experience this as our love of nature; or when it is threatening or hostile, as our dislike or fear of it. This remnant of primitive man remains in us — only today we feel we need to know why. The first spasm of developed selfconsciousness is to wonder — to wonder why things happen as they do; and later, to wonder at in wisdom. Primitive man, in his one-derness, could not wonder.

To primitive man plant life was a constant feeling of background companionship, not unlike the comfort we may receive from background music. But often he felt suffocated by its ubiquitous presence and he sought the deserts. He was just as finely attuned to the rock or mineral environment; even more so in the beginning, as he had evolved physically from it. But as his being began evolving emotionally towards the moment of selfconsciousness, man increasingly sought the stimulus of vegetation. Trees to him were like water spouts and water even in plant-forms is the medium of incipient emotion or feeling. We are mostly composed of emotion, or, as science tells us, water. Even today, emotionally excitable man is driven to live in the heat to dry him out; and less emotional man seeks the cold and damp to keep him supple. Deep space, or cosmic consciousness, towards which man is heading, contains very little heat, water or emotion. The consciousness of deep space consists mostly of an energy, a quality, called love.

The dawn of selfconsciousness

On one particular day primitive man watched the sun rise in the East. No man before him had ever seen the sun rising. He was never to be the same again. As he watched, cosmic culture was embodied in him. And the birth of selfconscious man began.

Sunlight is more than the energy that has been measured and evaluated across the existing scientific spectrum. It has a finer and higher psychic spectrum. Sunlight contains the initiating force behind selfconsciousness.

The sun had shone on the earth for immeasurable time, its eastern origins and western demise often obscured by the clouds and mists of the cooling, settling planet. But primitive man had never consciously observed it. While existing without he lived within. The sun and all external existence had been utterly obscured from his perception by a membrane that is behind all animal eyes — a veil of opaqueness. On this day the veil of opaqueness cleared and he 'saw' the earth for the first time.

The sun's rays were electric, penetrating the muddied mists of man's unconscious through to his psychic double. From deep within there was the faint response of another kind of sun or light — Man, the principle of intelligence, the germ-potential of selfconsciousness. And so began man's long exteriorising journey to existence.

The drive to selfconsciousness awakened in man on that particular day had to have an external trigger — sunlight. Nothing can come into existence without an external cause or trigger. The power comes from without, the idea from within. This ensures a synchronous and perfectly ordered evolution between the earth and the rest of the universe. All causal power comes into the mind from the external cosmos. From an external viewpoint the power to bring man to selfconsciousness was seeded in him on that day, like a pearl is cultured in an oyster, by sunlight entering through the eyes.

Primitive man's body had been on the earth for many ages before this cosmic culturing occurred. He lived in various parts of the globe. Some of his later habitations have been discovered. These have created a fundamental problem for archaeologists and anthropologists. It is difficult for them to resolve intellectually the finer distinctions between man's first civilisations and his primitive habitations, often older but in their way just as organised. The difference between the habitations and the first civilisations represents the interval between the demise of primitive man and the advent of selfconscious creative man.

The physical organ that finally made this momentous cosmic culturing possible was the pineal gland. It gets its name from the Latin 'pinea', pine-cone, which describes its shape, and it lies in the deepest and most protected area of the human brain. The pineal gland is present in all animals with skulls. In some reptiles it has an eye-like structure and is called the pineal eye. The precise functions of the pineal gland are a scientific mystery. However it is considered to be the remnant of an important ancestral sense organ. Science has found that in humans it responds to daylight through a link from the eye to the brain, and that it influences the body's awareness of time, particularly the onset of sleep, ovulation and puberty. These scientific observations become extremely significant when considered against the original evolutionary function of the gland, described in this chapter.

Science has yet to discover that the mysterious pineal gland was the original seat of the consciousness of man. Eventually, when a suitable body was ready, that consciousness exteriorised through the pineal gland to become 'I', the intelligence reading these words.

As the senses of the species evolved, so did the pineal. Their development was parallel: the two were intimately linked, yet separated for millions of years until the day just described when

23

the pineal gland provided the means for the ultimate superior sense, creative intelligence, to start coming forward.

The appearance of the pineal gland in the vertebrates coincided with the appearance of the first of the superior senses — sound or vibration. What I call the superior senses are senses like hearing, seeing and smelling that give warning-from-a-distance as opposed to taste and touch-feeling which provide no interval of warning.

When amphibians first crawled out of the seas they became test organs for the sense of sound. Both they and their descendants, the reptiles, had the pineal gland. It was the bony vertebrate structure of the first amphibians and later of the reptiles, pressed full-length against the earth, that gave rise to the sense of hearing. Vibrations transmitted through the ground, instead of through water as before, eventually became a sensitivity, still in the bone, to vibrations coming through the air — hearing.

The increased warning-distance which creatures gained from each of the developing superior senses equalled the evolutionary distance travelled towards selfconsciousness as the appearance of creative man. Selfconscious development was a growing sense of separation or distance from the environment or perceived events, giving time for reflection. Yet the early-warning system of the superior senses is not enough to justify the term 'selfconscious'. All the higher animals have the pineal gland and superior senses but cannot be called selfconscious.

The difference between man and animal, at all times and in all ages that have been and are yet to come, lies in the veil of opaqueness. This is a psychic membrane stretched across the back of all optical organs that have not yet evolved to that peculiarity of excellence known as the human eye. The psychic membrane is not an abstract notion. As soon as science concentrates seriously on finding a method for measuring the energies of the psychic spectrum — perhaps it will be a breakthrough of the new century — the membrane will be easily located.

The human eye, in spite of sharper vision elsewhere in the animal kingdom, is distinct among all eyes in having no psychic membrane. This gives man's vision a unique reflective factor which increases enormously the depth or significance of what he sees; hence cognition and perception. Freed of the membrane, man is able to reflect his awareness as perceptions back through the psychic double to the intellect behind. This reflective depth peculiar to man determines two things: first, the power of the intellect related to the sense-perceived world; and second, depending on the clarity of his psychic double through which the reflection must pass, the power of insight into psychic, spiritual and occult realities.

In the evolution of all the species, the veil of opaqueness and the pineal gland were intimately, vitally, connected. The function of the veil was to shield the pineal from the effects of sunlight until such time as a suitable potential animal body had been evolved on earth for I, the intelligence in man. All the necessary rudimentary senses, the anti-gravity devices such as limbs, walking and a co-ordinating brain, had to be evolved first in the lower species before they could come together for final development in the one form that would be physically capable of sustaining selfconscious creative thought, action and control.

In the evolution towards selfconsciousness, the I of intelligence seated deep behind the veil of opaqueness and the psychic double, had no sense of time. Nor did the exterior species. There was no way the psychic brain could know that a suitable physical body was ready for I to start occupying it, without some kind of energetic signal being received. The crucial signal to the psychic brain that such a body was at last ready for further development and for gradual occupation by I came when sunlight got through to the pineal gland. The pineal gland was the finest of all the earth sensors connected to the psychic network. For scores of millions of years it had lain virtually dormant in the skulls of developing creatures, not unlike a sensitive fire-alarm requiring a certain intensity of heat to set it off. All that stood in the way was the psychic membrane, which could also be called a membrane of forgetfulness.

So the vanishing of the animal veil in man allowed the energy of sunlight to stream through the optics and trigger the pineal gland. The pineal transmitted the signal back through the psychic double to the intelligence within. At once I, creative man, set out, up through the senses to eventually take his place among the rest of the species on earth.

For primitive man living his interiorised existence at the centre of the psychic mass, this meant the excruciating task of externalising his consciousness so that he could realise or relate to his physical senses and one day live in the world as we do today. It was an unimaginably traumatic, painful and time-extruding process. Primitive man had to be turned inside out emotionally and psychologically. From a completely subjective and undifferentiated experience of the world — something like the warm, comfortable, uncomplicated existence of the nine-month child in the mother's womb — he had to be made to go outside himself, to realise the greater truth of what he had become as the son of the earth. He had to be born, or re-born. He had to see, hear, smell, breathe, groan; and know he had a self, a body, that was separated from all the other things by a world he would later call creation, and later still, time and space.

Man first became selfconscious around 200,000 years ago. But only 10,000 years or so ago did the majority of men complete this extraordinary evolutionary contortion. Even then not all had completed it. And even today some still have not. The autistic and those with Down's syndrome are still trying.

A JOURNEY TO THE LIGHT

The story so far: before becoming a selfconscious being, man led a completely interiorised existence deep in the unconscious behind the senses at the centre of a vast psychic network or brain. He had no experience of his physical body which like the bodies of all the other species on the earth was evolving without any selfconscious presence or effort. He was linked inwardly to all things as a psychic or feeling ethos. He was like our brain which is the centre of our afferent and efferent nervous responses. What thought is to us, feeling was to him. He was the optimum of clairvoyant and extra-sensory communication except that he had no senses, no individuality, no personalised life.

On one remarkable day the veil of opaqueness had cleared from behind his animal eyes as he witnessed the sun rise for the first time. For the first time sunlight streamed into his unconscious via the pineal gland. He was now primitive cognitive man. Excited by the solar energy, man's psychic mass began to swell and curl into a powerful wave. Man was no longer content to remain a part of the unconscious. For the first time darkness was dissonance. A growing restlessness, vague stirrings, urged him up towards the source of light. He had no idea where he was going, only the feeling that he must go, break away from the inner mother, the now-stifling comfort of the unconscious womb, and find what he must — the truth behind his discontent.

Outside in the light, on the earth, the truth awaiting him was his physical body, unconscious and unknowing but fully functioning and prepared to receive him the moment he could manage that enormous journey to 'come to his senses'.

Two

THE GODS

*The truth behind the gods
of myth and legend*

THE GODS

FOR THOSE INDIVIDUALS who appreciate ancient myth and scripture, the ideas in them, like those in this book, have an elusively familiar ring of truth, like the feeling of a dream, which though distinct, persistently escapes the waking memory. Art, in fact, is the continuous endeavour to convey these original feelings ever more distinctly and meaningfully. From the mystical point of view they represent the indefinable quality called faith.

It is not the masterliness of ancient writings that has made them so revered and studied down through the ages. Nor is it attachment to their value as original works; they were set down long after the events they supposedly describe. Nor, as some may claim, is it because of their unique imaginative power: innumerable equally striking and richly imaginative works have been written since. What it is about the myths that penetrates and resonates to the depths of man's vital being (without his necessarily realising it) is the unconscious recognition that the events they describe represent the fantastic conditions actually existing at the beginning of the human race.

The sense-perceived world is organised in such a way that every stage of man's past evolution as a race is represented, literally recorded, in the phenomena around him. Consequently the scientist is able to collect solid fossilised evidence preserved in the earth and can gradually assemble a rational model of man's animal evolution. But this ordering of things does not apply only to biological and physical evidence. The evidence of man's creative beginnings is also there, waiting to be discerned in the same vivid way as the significance of the earth's fossils was once suddenly conceived. This evidence has been staring humanity in the face for thousands of years. It is contained in the myths and scriptural traditions of all cultures.

The Greek myths and Hindu traditions tell of gods and demigods walking the earth, fraternising with mortal man and even mating with him. Yet no one in the West today has ever seriously suggested that this is precisely how it was and had a hearing. Paradoxically, our civilisation has retained these scriptural and mythological fairy tales as an essential part of its reservoir of learning and thinking while regarding the essence of the myths — those earthly antics of gods and demi-gods — as fiction. And the accounts of gods and divinities are shrugged off, at best as allegory or some sort of symbolism. Even that wonderfully perceptive man, C G Jung, saw the energies of the myths acting outwardly through the unconscious but still failed to get back behind the symbols to identify the actual event in time that originated them.

The myths may indeed symbolise ever-recurring patterns of human and emotional behaviour, as Jung demonstrated. But that merely shows their importance as an effect, not their originating cause. What Jung regarded as fundamental is really secondary. Originating cause requires an originating event or an originating phenomenon. Nothing can have an effect that has not already happened.

Neither the Greek myths, nor the Bible, the early Sanskrit texts or any other of the most ancient religious traditions say that the accounts do not stand for exactly what they say. For us to assume otherwise is as arrogant and presumptuous as it would be if future man, looking back on the tattered records of our civilisation, dismissed evolution theory as a symbolic statement. Evolution theory means exactly what it says, only it happens by omission to be half-wrong; as the myths and scriptures happen to be half-right.

As the original myths were presumably repeated by word of mouth until written down, one has to expect distortion and imaginative accretions. Each race of people had their own traditions of gods, and as one group conquered another all the accounts commingled into the myths and legends we have today. These contain all the clues we need to explain the mystery of the gods — and the not-so-fictional first creative men on earth. And central to the myths and all religious traditions is the one point beyond embellishment that makes them such incredible tales in our experience — the unqualified reference to gods and demi-gods with miraculous powers participating in human affairs.

Were the gods — both mythical and religious — fact or nonsense? This question is at the crux of the whole science-or-religion dilemma, and the search for our origins. On what grounds can we venture to presume that the reports are not fact essentially? Common sense? Our common sense can tell us only what is commonly experienced now. It cannot tell us what the common experience was when man was actually becoming man, as the myths and scriptures try to describe.

The Hindu religion, one of the world's oldest traditions, speaks of the gods Krishna, Rama and other celestials walking the earth in human form. The same idea is essential to Greek and all mythology. The evidence of the myths is clear. The gods walked the earth and mated with man. Could anything be more straightforward and down-to-earth? The only question, it seems, is how?

33

THE GODS ARRIVE

On the world's first wave of self or emotion, man struggled up the solar stream of consciousness towards the surface of himself and the earth. Leaving his interior existence in the psyche, he began to consciously enter and realise the physical body prepared for him.

Several men made it first. It was not unlike the arrival of the first few swallows in spring. For a long, long time these first physically conscious men were a very rare phenomenon in the small communities of unselfconscious men who lived together in various parts of the globe.

As each selfconscious man came to his senses to perceive the wonders of the earth he was also able to retreat back into the psyche. There he could communicate to those coming up behind him vague notions of the amazing existence awaiting them in body-consciousness. This was a very simple thing to do. The first selfconscious men were so loosely attached to their physical bodies, compared to us today, that they lived in both worlds. They had only to meditate, to withdraw from their external attachments and physical senses, and they were back in the instinctive, inner world where all men were one. These first-men could take back with them their selfconscious identities consonant with their physical realisation. They were able to inject into the instinctive network of all potential rising men not only the powerful energies and feelings resulting from their sensible awareness, but also the extraordinary practical knowledge it gave them. To the mass of unselfconscious men these literally were beings from outer space or another world.

All men approaching body-consciousness were at different levels of advancement or emotional evolution. Some were very close to physical realisation. Others were thousands of years away from even the possibility. Between the two were innumerable 'men' of all shades or degrees of incipient self-consciousness.

34

In the outer physical world all the men living together in communities looked very much the same. But those with more acute comprehension of their physical existence were aware of the differences among those with inferior comprehension; and they took advantage of it. They used this knowledge to exploit their inferiors, so laying the foundation of the later class systems.

A physically realised man was a very privileged specimen. He enjoyed tremendous advantages. He could communicate with his developing companions in either the outer world or the inner world, on the earth or in the psyche. While in the company of lesser men on earth, he could tune into their psychic state and know exactly what each one was feeling, what he wanted at that moment and what he would do next. He could speak inwardly to the lesser man like some unidentifiable, all-knowing presence. He could put suggestions into him, frighten him, persuade him, instruct him; and finally, he could induce him to worship and serve him. These first physically realised men were the mythical and legendary gods.

Nothing was impossible for the gods. With access to both physical and psychic worlds or realities, they were able to create in the psyche the monsters, visions and all the vague threatening demons which imagination can convey even today. But it would not be correct to describe theirs as a fantasy world. It was *all* reality.

For many thousands of years, while man was becoming physical and entrenching himself in his senses, the world consisted of the primitive psychic as well as the physical material. No rational line could be drawn between fantasy and physical reality. Men lived in both worlds but did not know it; the gods, the realised men, also lived in both worlds, knew it and had the best of both.

The gods were men. They mated among themselves. They hated and loved and warred among themselves. And when they desired a shapely unrealised man or woman's body, they took it.

35

THE GODS' GAME OF LOVE

Now it is possible to understand the origins of the myths of creation. The mystery, the delusion, the endless intellectual theorising and speculation start to fall away, revealing that the truth is even more fantastic than the legend.

In ancient Greece a community of realised men, or gods, lived on Mount Olympus, or so they made out. And the myths say that the Olympian gods loved many mortal women, especially Zeus who is described as 'omnipotent king of gods, father of men' (unrealised men) and master even of fate. And why not? The men who handed down the myths were not gods. They were the semi-conscious masses, whose hazy psyches were used by the gods to wage fantastic wars against each other with thunderbolts and other earth-observed phenomena, crudely scripted and sculpted into demonstrations of illusionary power and magic.

Zeus fathered four of the Olympian gods by mortal women. Why not? His body was as mortal as theirs. The difference was that Zeus and the other first-realised men and women never died. When their bodies wore out they withdrew into the psyche and took possession of another suitable body. If they chose, they could even push aside a man coming to conscious birth for the first time and take his body.

The gods could scare the daylights out of their ascending 'subjects' by swallowing children, as Cronus did, and creating gigantic monsters like Typhon whose huge limbs ended in serpent heads and whose eyes breathed fire; and then they could destroy their creations and bury them under Mount Etna, which still 'breathes fire'. To escape each other's ghastly, ghostly creations they could turn themselves into rams, crows, goats, cows, cats, fish, boars, ibises — anything they fancied. Goddesses like Pasiphae could mate with magnificent white bulls and in a phantasmagoria of erotic and sexual imagery conceive monsters such as the Minotaur which fed on maidens and young men.

The gods of Olympus were a group of particularly privileged

men waging wars of dreams among themselves in the primal mind stuff of a humanity that could only look on in stupefied awe and terror.

The physical reality, however, was very different. The gods, living in a world of limited physical comfort, dreary, mundane company and endless time for reverie were unable to desist from the game of creating in the psychic plasma. Here the audience was live and unbelievably, feelingly real, offering a response of concentrated awe and wonder which every artist, entertainer and illusionist has dreamed of creating ever since. By comparison with their psychic artistry and games-playing, the physical world was impossibly slow and tedious, the audience nothing more than servile zombies.

The physical sex of the gods and goddesses was good, but never enough. How could it ever be enough? Never could the senses deliver in the final consummating moment of orgasm a potency of feeling equal to the tantalising excitement and delirium of their extended psychic fantasising. Today's imagination is a wraith beside it.

In their physical sex the gods and goddesses were able to communicate to each other, or to their lesser mortal paramours, the full reality of their erotic dreams. Each participant literally became any role they were playing while both or all shared in the living experience. Which was real — the drab physical mating action or the endless psychic stimulation? Both, but inevitably the copulation ended in the drab physical reality. 'I'm coming' was the cry in every ecstatic climactic feeling and has been in every language ever since. Every lover, man and woman, eventually comes to their senses, back into the world of the living.

The gods were men of the most extraordinary consciousness. Yet it was not that their consciousness was any more profound than man's today, only that it was not burdened with past. Having no past as men they could do what they wanted and be what they were — free as no man or group ever again would be. Past is the constraint which dictates the future, the inevitable.

As these men had no past they had no future. They could exist only once, in the beginning, never again.

The age of the gods was relatively short. Every moment they spent in external awareness past was building up within. Finally this pushed them back out of sight and over the horizon of time.

For the waves of men coming to existence behind them, the creative and imaginative past initiated by the gods with their psychic games-playing would be an ever-increasing burden in consciousness. What the gods were and what they did had to determine for all time the future and nature of the human race. This legacy, in which all men share, is sexual pre-occupation, erotic fantasising, creative imagination, the longing to regain original freedom and divinity, the use of illusion to obtain power over other men; and the cultural legacy of the myths themselves that tell this story.

Nothing reveals the humanity of the gods — the future they set for mankind — more than the emphasis they put on sex and the inevitable crudity of those concepts. They had emerged from a world in which the sexual act did not exist into a world where it represented the heights of love and lust. The psychic world is a world of fluid being, of attraction, repulsion and embryonic love. In the psychic world energy such as sex cannot build up; it is immediately and continuously released and dissipated throughout the entire psychic plasma. For the gods, entering into physical existence meant that their experience of the psychic flow was to an extent blocked by the senses. For the first time they became aware of the build-up of feelings of sexual desire which could be released, or heightened and prolonged, in the sexual act.

Sex is a physical experience, but actually a psychic reality. Sex as we know it cannot be performed in the psychic world. But it is felt there. As we know it, sex culminates in an orgasm, in a glandular climax. It is a sort of implosion, the release or return of a particular, pent-up psychic force. All physical experience is made up of a moment-to-moment, bit-by-bit damming of the continuous flow of psychic energy. This damming actually creates our solid physical form and the procreative nature of the genital organs.

But the energy of the genitals is not for all time to be procreative, in the sense of merely producing physical beings. Behind the male and female reproductive organs are two intensely creative though at present largely separative energies — man and woman. As future man and woman make the psychic or vital world a conscious part of their existence by being made to learn to truly love, their energies will merge and harmonise without loss of individual consciousness. Man on earth will then be united with the one divine woman all men seek, and woman with her divine man. Together they will create a new psycho-physical race of people on earth. From birth their living experience will be of the psychic reality as well as the physical world giving them an intensified sense of purpose and responsibility for life in both worlds. This future wave of men and women will be a new, mature version of the gods.

Three

HUMAN NATURE

*This is the story of human nature
and how it forms as a substitute
for man's lost immortality*

HUMAN NATURE

THE END OF EDEN

THE GODS HAD gone, vanished forever, leaving behind the first gentle swell of past that began with them. The turbulent transition of selfconscious man to earth and flesh was over.

A short period of extraordinary tranquillity followed. Through his new-found senses man began to participate in and enjoy the astonishing nature of the earth and to breathe into his being for all generations to come the living memory of its richness and beauty. Mother Nature embraced her newest child.

At this time, when man and his past were very young, nature was the only existing order of things. Nature was the earth's immemorial past, consisting of yesterday's experience in plant, animal and man being applied as life today. There was no planning, no future — only action and doing now. As a part of nature, man had no expectations and demands outside of what it was and what it provided. He had no desire to change it any more than did the rest of nature. Animals, birds, trees, flowers and men: what had existed yesterday disappeared back into the earth while all was continuously, effortlessly, miraculously replenished today. Nature was complete — the natural past unfolding as the natural present.

The gentle pressure of his own young past kept man finely balanced within his senses. He used the senses in the way nature had evolved them, as receivers, permitting him to stand back within and relish the sounds, smells, sights, taste and touch of the all-pervading earth. It was a wonderful, simple state of living and being, of yesterday and today. Tomorrow never came.

But one day tomorrow did come. Man invented it.

The halcyon period in which man was in his element, his Eden, ended when the past the gods had started but never shared in, at last caught up with him. It came in the form of death.

Instead of accepting the natural cycle of forms disappearing into the earth, man started to register their disappearance as death. He was becoming attached to his external senses and losing his knowledge of the inner immortal world. His new perception meant he was starting to believe his senses; he was attaching himself to the appearance of things. This made no difference to the truth that there is no death and that the energies of all things that have lived survive and move instinctively to live again.

The surviving selfconscious energies of the first men who died came back to earth to live again. Behind them and latching on to them were streams of instinctive energies vying for the experience that only imaginative, selfconscious man could provide. The gentle, natural swell of returning life suddenly had become a huge wave, an irresistible surge of yesterdays thrusting into the most sensitive point in nature, the human brain, to emerge and live again. The mounting pressure drove man further and further into his senses, away from his inner nature and towards the formal world.

Finally, the pressure from within started driving man out of his senses. He began projecting through them into the world. Rising emotion started to appear in his eyes and gestures. His sense of smell reversed from a soft intake of breath to an often harsh and aimless projection of words. His sense of taste became a missile of opinions often expressed as violent likes and dislikes. Hearing developed into an emotional extension of his body, alert not for natural sounds but for intonations of insult and slight.

Attachment to his body increased. The further out he projected, the more isolated, vulnerable and threatened he felt. This in turn made him cling emotionally to his body even more. The earlier fine balance between nature and the senses, his original inner poise, was destroyed.

Man had extended himself so far that he was now outside of nature. Nature, like himself, had become for him a purely external sense-perceived form. It could no longer provide him with that intimate feeling of oneness with the earth and all life.

In his new exposed, externalised position, all he could see in the environment was the threat of death. To survive seemed the only realistic point of living; yet survival was impossible, as the evidence of each day's dead bodies around him demonstrated. Death, once the most natural event, was now the most terrifying. He became obsessed with the fear of dying and losing his body, the last apparent formal link with where he had come from. He started to see each dead body as an extension or projection of the end of himself.

The now formalised beauty of the earth was as nothing under the pressure of this awful outward-going fixation. Man was beside himself. He was being consumed by his own psychosis, his emotional attempt to limit life and today to form. He had reached the end of his tether. He could go no further into the formal world without losing touch with his senses altogether. He had to find a way out, an escape, some other form of today, another world that did not end in death. Man was on the verge of insanity.

Tomorrow was his answer. With this one stupendous invention he could break out of today and create a new world of his own, a world of progress and continuity, an apparent ongoing, a distance between himself and death, the end. Death could now be forgotten. It would come tomorrow — never today. In this way it became a problem that no longer needed to be confronted.

This was the world's first act of ignorance. The truth is immortality — man's original nature, which he had allowed himself to be driven out of; and any substitute for it is ignorance.

Tomorrow is so naturally a part of life today that it is almost inconceivable to us that once it did not exist. But our difficulty is only the inevitable result of that same ignorance which is now at the foundation of human nature. As with the behaviour of the gods before him, man's means of escape from the truth was to determine for all time the fundamental condition of his new human nature, as well as the inevitable course his projected world of tomorrow and progress would have to take.

By identifying himself with the idea of world progress and improvement, man also unconsciously invented the necessary delusion of hope. Hope is not necessary for today; it arises only in trying to escape from today. In a world where man must die there can be no hope, apart from the delusive hope that he will not die. Hope is an attempt through ignorance to put off the inevitable.

Instead of living naturally for today man began living unnaturally for tomorrow. This meant projecting himself emotionally into the future; that is, further beyond his senses and the formal world into imagination. But the more he looked to the future in his imagination, the more dissatisfied and discontented he was with today. This impelled him to start imposing a new order, his own idea of how things should be tomorrow, on the natural order of how things were today. Then, by constantly looking to see how things should be, he started to lose sight of how things were.

Dissatisfaction and discontent, plus the peculiar ability to think straight about everything except what was most important, were added to ignorance at the foundation of his new human nature. Man reeled from reaction to reaction, every reaction contributing to a delta of more effects.

MAN PROJECTS HIS NEW WORLD

In those days when the past was young and man was still sub-consciously looking for an identity to reflect his forgotten immortality, any idea he could identify with became part of the new human nature he was creating for himself. He was still fashioning and ordering the human psyche and its human way of perceiving.

Perhaps, in retrospect, the forming of human nature — man's substitute nature — may seem a very profound and complex process. But in fact it was no more than the complicated outcome of a very simple psychological device for escaping from emotional pressure. By creating innumerable problems around a supposed future or tomorrow, man was able to keep his mind busy and distracted from having to face up to the one big problem of today — death, the end of any future or imagined tomorrows.

This psychotic nature can be seen operating in every man's worries today. To worry today he has first to assume he is not going to die today. If he were truly aware he was dying, or going to die, all his other worries would vanish instantly, like the chimeras they are. Those who are dying continue to try to tell him this in their late-found wisdom. But ignorance, the hope of an imagined tomorrow, closes his comprehension.

So death was projected hopefully as an event in the future. And by all men projecting the same idea together in the same hopeful, emotional and ignorant way, death ceased to be a necessary consideration as a threat or end to the progress of the whole. The individual might die; but the world of tomorrow would live on in the hopes and imagination of all other men. For all generations to come men could talk about death, plan around it for the future, even joke about it, without ever having seriously to examine it or feel the least bit menaced by it. Hey presto — death, the only real problem in life today, was eliminated.

In those days man's past was still young enough for him to be born with individual consciousness; that is with the knowledge of immortality intact. But under the new order of time and ignorance it was essential that this individuality be eliminated. If allowed to mature, such self-knowledge of immortality could destroy the delusion of tomorrow and perhaps even human nature itself; as well as all the sophisticated and progressive men who depend on the idea of the world for their existence. Such individuality would possess the power to start turning back the wave of progress and return man to the point of sanity where tomorrow and the world no longer really exist and the past no longer matters. This would be intolerable.

To keep humanity from such dangerous self-knowledge the subconscious of every child was instilled from birth with the idea that this is a world which goes on and on, irrespective of what happens to anyone. Tomorrow is all that counts. To question death and dwell on it is morbid and unhealthy. People die but it does not matter because tomorrow is a new day and death can be forgotten by everyone in the promise of the future.

A remarkable change now occurred in the human psyche. Whereas man had once despaired to see the end of himself and everything else reflected in dead bodies around him, now in the bodies of his new-born young he saw a splendid living reflection and extension of himself, as well as the hope of the world of tomorrow. When a child was born he proudly rejoiced, and the whole world with him: not for the new-born immortal consciousness but for the limited, mortal form that the individuality (with man's ignorant connivance), would be forced to identify with in the world of tomorrow.

Every child from birth would now be immediately and ceaselessly labelled, first by his parents, guardians and teachers and later, as he got the idea, automatically by himself and the world until he fervently believed in both. Initially he would be labelled 'baby' and treated like one; taught to comprehend, think and react in an infantile way, matching the ignorance of his parents and the people around him. Never would the opportunity be taken to intelligently address the immortal, unperturbed and clearly non-conforming individuality in the new-born body. The parental duty from birth would be to immediately assume the protective position of the teacher of ignorance and mortality.

This was man's ultimate projection and his supreme act of selfishness. In his ignorance and for all generations to follow he would sacrifice the uniqueness of his new-born offspring for his conforming idea of the world, the future and himself. This he perpetuates today in the form of parental and filial love.

The rest of the labels identifying the child as a progressively doomed idea in the world follow automatically: one-year-old, two-year-old, three-year-old, boy or girl, good or bad, teenager, student or layabout, man or woman, worker, manager, managing director, artist, poor or wealthy, happy or unhappy, mother or father, grandparent, old, boring, senile, dying.

'Dying'. When finally this last label was affixed, the idea of progress would vanish for the man after a brief reappearance of his individuality towards the end. Birth and death, the entrance and exit of man's progressive world, were the only places left for the appearance of individuality. In their grief, those closest to a dead person might have some temporary doubt about the sanity of this world or themselves, or the inferred security of either. But such anxieties would soon be obscured by the compulsive delusion of hope and tomorrow. The whole delusion depends on the hysteria of mass forgetfulness: never rooting out the cause today but looking to see how it can be removed in the future.

In man's world of progress all individuals are expendable because they have no place in it. Since the individual is the only one who dies and the world does not recognise death, dead individuals are remembered as figures, as the positions they filled, and are forgotten as men. In a world of progress and tomorrow a dead man cannot be missed for long because his position in such a world was only an imagined one anyway, and will promptly be filled by the imagination of the living. There are no spaces and no space in man's world.

EVOLUTION BY INVOLUTION

All of this, the whole saga of 'tomorrow' and the forming of man's human nature, was the beginning of the process of evolution by involution.

Up to the point of selfconsciousness, evolution had been confined to the physical ascent of the species — or the descent of man into matter, his body. The next phase of evolution — involution — was to be the gradual enforced descent of his self-consciousness into a new body, a world body of progress, rationality and information, later to become today's intellectual materialism; a world of no real knowledge since it meant leaving behind his knowledge of immortality.

Involution is the endless complication of life by man's self-consciousness. This process requires him to go further away from his immortality and deeper into the world-body of ignorance, falsehood and imagination, before he can get back out and free of it. Due to the emotional escape into tomorrow, evolution by involution is paradoxically man's only way back to immortality and godhead.

Man invented tomorrow and now has to live in it. As he determined the future when his past was very young is how it must be — except that finally the truth of his immortality and godhead must out. He must return to what he was. By going out he comes back.

Evolution by involution can be illustrated by the analogy of the wheel, a familiar symbol of time and progress. As the axis of the wheel of evolution moves forward, the wheel itself revolves backwards over old ground.

Today, man is still engrossed in constructing for himself his new body, a world-body of informational knowledge, just as Man, the principle, constructed his physical body. The world's stupefying daily production and exchange of information as news, opinions and data suggests that this immense body-building exercise is well on schedule.

The individual man must become thoroughly entrenched in

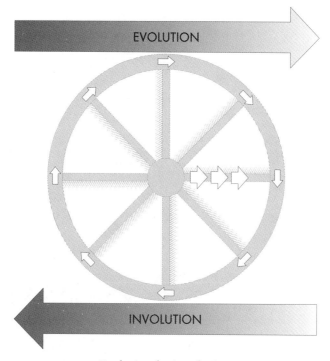

Evolution by involution

this new body (his 'knowing habit') before he can begin the task
of freeing himself from it. This he does by realising in himself I
the intelligence that supports every form of knowledge and
body-consciousness. Then and only then can the next stage of
evolution begin: man's voluntary re-entry into the psychic world,
this time through self-knowledge and as a fully conscious and
no longer selfconscious or self-orientated human being.

Four

VIOLENCE AND CIVILISATION

*Each life is lived solely for
another life*

VIOLENCE AND CIVILISATION

THE FIRST SOCIETY

WHEN TWO OR more selfconscious beings are in each other's company a powerful psychic atmosphere is generated between them. It distinguishes man from all the other animals and is taken completely for granted. This atmosphere is charged with man's emotional idea of the world as progress, continuity and tomorrow.

One man in isolation from his fellows for long is in danger of losing touch with the world, or the idea of it. Two men in contact together in isolation, even though competing against each other for survival, will keep the idea of the world alive between them. This is the answer to the riddle of the wolf-children: would a human baby brought up by animals in the wild remain human? Because of the unique humanising atmosphere generated between selfconscious beings, two children raised together by wolves in the wild have a chance of remaining human, or in touch with man's idea of the world and tomorrow; but one raised alone will remain virtually an animal. Conversely, animals such as domestic pets in the constant presence of human beings are seen to become 'intelligent' or more human. They pick up the progressive idea and accordingly tend to act less wild. Wildness is merely the state which refuses to accept or conform to man's idea of the world.

The first selfconscious men to generate this uniquely human atmosphere saw — or in the presence of each other thought they saw — that the world, the things and people around them from which they had now become largely separate as self-entities, was not perfect. They inferred that things needed changing, that an order apart from the natural was essential. They were getting the progressive idea. Simultaneously, society began to form.

A society is not a habitation. A habitation is where people live together. A society is where people live and work together linked by a common ethic — the idea of progress — under some kind of law or set of rules. Society is the forerunner of civilisation and the need for progressive order. Civilisation is the need for pleasure and art and these, man's most civilising drives, are his subconscious attempt to recreate the world of the gods representing the ultimate in pleasure and artistic creation.

Man's first society was shaped by the sheer necessity of his new selfconscious existence. Society had to follow the exploitative example set by the gods, as well as man's own escapist idea of a collective existence whose continuity must never be interrupted by the serious consideration of death.

Death, in effect, was outlawed as a proper topic of conversation and particularly as something to ruminate on. Death was to be buried — never left around to disturb or disintegrate the progressive social idea. The show must go on. If death had to be seen or acknowledged it must be dressed up, disguised and given some sort of collective dignity to make it part of progress, a sacrifice, an on-going, a ritual, a tragedy, a rest, a wake — never faced as its sordid, hideous self.

Man was now somebody. He had both an individual and a collective or progressive identity. He had something to live for today and to die for tomorrow along with his new compulsion to change the world. But the message inherent in the evolution of selfconsciousness (as distinct from its involution) is that to change the world (and not merely to move things or himself about) man must first change himself. Already few men could understand what this change was; and even fewer were — or now are — prepared to make the change.

The largest class or type of man found self-change impossible. He had to protect and raise his young and to do that he had to make sure of his survival as he was. Changing himself would have to wait. This was the family man. While he was busy trying to survive and bring up his family, he found he had neither time nor adequate means to protect them. So he offered another type of man in the society food and shelter if he would fight off his enemies for him. The warrior caste emerged. The simple and more stupid kind of men, for their keep, became the servants and menial class. And, for the dirtiest jobs of all, the insane, crippled and diseased were used — the wretched untouchables.

The last and smallest class of man had very different ideas. This class saw that a celibate man did not need to worry about surviving to protect his offspring. He could concentrate on finding the power within to change himself, and so perhaps eventually change the world. This was the self-searcher, the original monk (from the Greek 'monachos' — alone).

For man who is celibate by choice the company of his fellow man is not so important. If he is looking for something within himself he is inclined to live in isolation or seclusion, away from external distractions and the idea of progress. A man trying to find the power within himself is not so concerned with living or dying: so, many monks wandered away and died of hunger; or they were killed by wild animals or other men.

Some of the monks who survived realised the search for the power within — and they regained the idea of their immortality. Others pretended to. The pretenders, and those who had only partial success were the first priests. They entered the habitations of men and set about distorting the truth. As part of the involution of the evolutionary movement towards all men reaching the truth, the priests corrupted it with rites, ceremonies and interpretations, and with magic or psychic force out of which arose worship and superstition.

The monks who did realise the truth fell into two categories. The first and larger group of self-realised men, although very small in number, were those who had realised the simple truth of I, the immortal intelligence — a realisation which annihilates the need to change or search for anything. These holy men wandered or settled, drawing to them monks and other men to whom they taught the truth. The other, even smaller category of holy men emerged from among the first group. They were the original gnostics. These first-wave gnostics and their successors had realised not only I but also that level of mind which reveals the purpose of life on earth. This realisation reawakens the knowledge of mankind's long-forgotten origins — gnosis.

THE ORIGINAL GNOSTICS

Monasteries and their equivalents were set up by both priests and self-realised men, but it was the gnostic masters who founded the first real spiritual centres: 'real' in the sense that they were actual working models, demonstrating the means and purpose of terrestrial life, or the gnostic realisation.

The gnostic master knew, through the knowledge imparted by his realisation, that for him to attract right men to build and work in his centre, he had only to sit and wait. He had no sense of impatience. He could wait forever. If his body died, he would come again, or be again. Once realised on the earth he was always there, custodian of the gnosis which would be realised afresh in each life-form, for as long as it was necessary.

The original gnostics were already in existence when self-consciousness was very new. Each of them lived to an age of several thousand years. This was made possible by two factors. First, in those days the organs of the body could still be renewed or revitalised from within through the psychic double. Second, the original gnostic masters were able to occupy a succession of vital and physical bodies surrendered to them in selfless love by devoted followers.

At that time the physical constitution of human bodies was different; they were less solid in the sense that the psychic double was less entrenched in the physical habit. (In this lies the origin of the monk's habit, his traditional tunic.) And further, the degree of individuality was variable: the psychic double (the reality behind the word 'soul'), alternated between the two worlds — contracting to appear as the ego in the physical, and then expanding back again into the vital. Sometimes a man would be fully physically orientated; at other times he would be mainly in the psychic world. After a time, as involution increased and man became more exclusively absorbed in the play of sense-perception and the physical habit, the double flowed more rapidly backwards and forwards between the two worlds, finally to become our rhythmic breathing principle. (The true significance of the Greek word 'psyche', meaning both soul and breath, will be appreciated from this.)

Each gnostic master formed his spiritual centre around himself. The physical centre was built in a remote place, well away from other centres or human societies. The masters made their centres fully self-supporting. Every man permitted to enter as a student had his worldly job to do as part of his spiritual preparation and instruction. Implicit in entrance was the unconditional surrender of his will to the master. He had no rights and abandoned all expectations, holding on only to love, faith or trust in the master.

The gnostic centres were compact, refined models of the first societies, except that the defence and control of each centre was completely in the hands of one man, the master. He personified fundamental justice: he knew the truth and he wanted nothing. The original gnostic masters repelled would-be intruders in the same way as they attracted right students — through the power of the Man they had realised within them. No hostile person or psychic force could penetrate their centre.

For us today this would indeed seem to be an ideal, a state beyond circumstantial possibility. But we are approaching the evolutionary point where for the individual the way back to it can begin. The future men who eventually become gnostics will be different from the first gnostics; they will have come to this state through selfconsciousness, exist in larger numbers, and have a far greater influence on the humanity of their day. Theirs will be the knowledge towards which humanity will be consciously aspiring — not heading away from, as was the case in the time of the original gnostics when evolution demanded man's descent into the matter of circumstantial existence, which is life today.

In the aeons that followed the founding of the original gnostic centres, involution — the process of complication of man's ideas and his world — began to grip. Gradually but inexorably, intellectual complication overlaid and diminished the power of succeeding generations of gnostic masters to control matter as themselves. The gnostic masters had been fighting a losing battle from the start. Under the weight of matter as concept and wanting, the loss of their power was inevitable. But their purpose had never been to retain the power, anyway — only to preserve the knowledge by teaching it and realising it. The power would look after itself; and reappear when the age of selfconsciousness approached its end.

THE MASTERS' POWER

The defence of the centres became more and more a matter of combined action by all living in them, although the emphasis continued to be on utilising the power in the individual. The later masters taught their students to deter attack by projecting the power without movement through their bodies in the psychic tension of confrontation. As this power also began to wane, the movements of the student's body were used to harness the attacking force and turn it against the attacker without initiating any violence. Then came the unarmed blow directed at the psychic centres of the attacker's body, demobilising or killing him. And finally weapons were introduced such as the sword and the bow, which the masters taught their students to use as psychic extensions of their own realised power, or vital force, within.

All these forms of meditative combat taught by the long succession of gnostic masters in the original long-forgotten East are represented in the martial arts that survive today.

As originally practised the martial arts were a means of generating power in the man. This was such a peculiar quality of power that although the master of the art possessed the power to kill or disable, he could never use it and keep it. To use the force was to lose the power. This paradox characterises the culture which man is now heading back towards. And this time he will have his enormous experience and understanding of life to draw upon.

The power is what Jesus of Nazareth was talking of in his doctrine of turning the other cheek. It is the power behind the teaching of surrender, essential in the Old Testament and in Islam (the word means 'surrender'). It is the power invoked or invited by the state of desireless action central to Buddhism, the power practised politically by Gandhi against the British in India, and the strength behind the doctrine of non-violence pursued by many great and effective men and women in history.

The extraordinary power latent in the individual man is something the world today has little understanding of, or time for. It has been perceived by many men but not realised by enough of them to make it a significant truth for all. At present this power is on the descending node of evolutionary causes. Westernisation is in ascendance — the ideal of world progress powered by the involutionary drive into intellectual materialism. Consequently the power is not permitted by the times to be publicly demonstrated, except in rare instances. These events if reported, make the headlines as miraculous and mysterious happenings and are quickly forgotten, except by those who experience them. The westernisation of the masses demands continuous streams of information — not knowledge.

WESTERNISATION

The idea which puts faith in the power in man comes from the original geographical East which has long since vanished. The eastern martial arts we are familiar with today are one of its superficial expressions which have managed to survive the world-swamping wave of westernisation, the dive into the substitute world of intellectual materialism. Almost all of the East, even that remaining less than a thousand years ago, has now gone West.

The westernisation of the world is not a modern phenomenon; it has been going on since the start of involution. However the first significant crystallisation of the process came during the Roman Empire with the introduction of authority based on numbers and force. The next stage of westernisation, Christianity, added the emotional impulse necessary to bring us up to the recent present. Now westernisation is in full swing. Every nation on earth — backward, developing and developed — is striving for the same technological, material and intellectual ends. Individuality seems only to get in the way.

It is only a matter of time before the original eastern culture of faith in the power in the man begins to assert itself again. When it does, it will not be in the form of the ancient eastern traditions as they are popularly understood today. The new culture will manifest as the culture when the westernisation of humanity is completed — when finally man has come to his senses by realising what is of value. The force of the intellect will then become the power of intelligence: the ability to discern unseen causes, principles and purpose.

There is one massive problem. The birth or release of the new culture from the human unconscious, where the idea of it is retained, will require the destruction of most of the human race by nuclear war, holocaust or some tremendous natural cataclysm. This could happen any day now.

Why is it we hear about the fighting arts of the East — the martial arts — but not about the fighting arts of the West?

In three thousand bloody years of recorded combat, the only empty-handed skills the West has handed down to us are boxing and wrestling. Considering that long and violent opportunity for developing a fighting art-form, these skills remain crude and undistinguished. The West has made a business of fighting, a profession and a necessity of it. The developing western identity, the western way of life, has provided the world with ever-better weaponry or killing power. (The Chinese apparently thought their gunpowder invention was for making fireworks until western ingenuity demonstrated a more profitable and business-like use for it.) But the West has not made an art of combat. Not because its fighting men have been any less skilful, courageous and noble, or murderously determined. But because the emphasis has always been on faith in the power of the weapon, or group, and not in the power of the man, the individual.

Over the centuries, the West has gradually built up its power and identity from number; from numbers of weapons, carried by numbers of men, conglomerating into one single armed force or power pack. The ultimate effect of this today can be seen in the sheer numbers of men that can be destroyed with a single weapon or unit of force — the Bomb. All history's innumerable fighting men and their weapons have now been made redundant by this stupendous unit of annihilating force. The Bomb is a brilliant vindication of the western philosophy of putting faith in the power of the weapon — a fact indisputably measurable in its potential of dead men. It is the ultimate product of western rationalisation, the final solution to the problem of modern logistics, a superb example of modern economy. All those millions of soldiers it has supplanted lie like phantoms inside it, ready to manifest as millions of the dead and dying the Bomb will leave behind when it goes off with the collective force of all their weapons fired in one catastrophic salvo.

It is a marvellously efficient philosophy. But its nuclear brainchild is not art. Such handiwork could never be included in a list of the fighting arts. In the case of the martial arts, the western way of thinking would argue that a knife or gun would do the job more easily and quickly. Any killing or maiming, or threat at the point of a knife, gun or bomb, can never be the point of art.

If an exponent of the martial arts is to become a master he must eventually realise that even a threatening stance or hand cannot represent the art he is seeking. The power in man, whatever its quality, must be uniquely different to the threat of the weapon, the force of number or the fear of a blow. Any art has to be more than a skill. It must possess that extra something, a quality which penetrates and edifies the consciousness of the observer beyond the excitement and pleasure of being entertained. Art is for art's sake. It is not for winning.

The fighting arts of the East as popularly practised today do not have much to commend them as art-forms, except where meditative disciplines based on conscious restraint are demanded. Nonetheless, the people in both East and West seriously striving to master these martial arts have a great deal in common with the many more who seriously pursue the other less physical art-forms. Both these groups are drawn by the same attraction — the martial group being pulled towards the source of original power, and the creative group towards the source of original art. In both cases, the source is man's forgotten godhead.

MAN, THE FIGHTER

The fighting spirit is that noble thing which moves us when we see it in a game, contest or attempt to overcome adversity. Its opposite is the fighting nature. The fighting nature is ugly. We fear and distrust it. When we see it we hope to avoid it or see it vanquished. The two are the difference between a fair fight and a massacre, a skilled defence and a knife in the back, a brave man and a bully. The fighting spirit is in all of us. It is constantly trying to get the better of our fighting nature. We experience it as the effort to control ourselves when provoked. The fighting nature has no self-control. It lashes out with an angry word, a kick or a blow before the man in us has time to think. It is an instinctive reaction, a primitive, mindless impulse to hurt or disable.

This is not to suggest that everyone is going around looking for a fight. The problem of the fighting nature for most of us is more subtle than that. We are regularly assailed by hurtful words or critical attitudes. And, whether we admit it or not, most of us are just as ready for a fight or slight at the drop of an insult, just as quick to hurt; and to regret it after the damage is done.

The fighting spirit in adversity *is* man; but not the animal part of him, his fighting nature, which clawed its way up from the swamps and slime to meet him at the point of destiny where the human spirit entered matter. Man is separated from the animal, the primitive, cruel, survival instinct, by his ability to sense consequences before he acts. His body — tongue or fist — will lash out blindly and start chaotic sequences of violence. But not when his fighting spirit is developed. This prevents him from reacting blindly. For a split second he senses the consequences. There is a pause. This is the moment of truth: to do or not to do, to strike or not to strike, to kill or let live. It is this intelligent pause, cultivated in the swamp of our base instinct, which is the difference between the admirable response of the fighting spirit and the violent reaction of the fighting nature.

We are faced with the fact that man is violent by nature, and reaction. And as we look around we apprehend the rest of nature is violent too. Yet it all works: life is good. Life is hard, is filled with unexpected death and injury; is often sad, cruel and unfair, but it is good. How can we say it is good? Because every single one of us is fighting or living for something beyond survival.

One of life's mysteries and absurdities, however, is that most of us find it impossible to explain what this precious something is.

After all, it is no new discovery that each of us is going to die. One would think that with the death-sentence already pronounced we would be distraught, unable to go on. But most of us are not. For us not to be able to say why we keep going towards the day of extinction, and with such relish too, indicates that we are not facing up to something, or that there is more to our life than most of us are consciously aware of. Despite our exertions after bigger houses, new cars, love, art, drugs, transcendental experience, lovely women, good men, sons and daughters, fame, power, reputation, these things are never enough when we get them. Certainly they are part of the means to whatever we are living or fighting for. But obviously, they are not the end we seek. Yet, each day we fight or strive for more of them, or to keep those we have. Why?

'It's natural', we say, 'That's life'. Which is true enough but explains nothing. All that ceaseless trying and struggling which the human race engages in is a means of absorbing and occupying the tremendous massed force of our fighting nature. If we were not made to expend the bulk of our energies in regulated, acquisitive directions we would all be busy killing each other.

The fighting nature of humanity has a similar diabolical potential to the Bomb. Our efforts to possess men, women, experience and power, under the almost as violent protection of man-made law, makes living tolerably safe while we get on with it. Each day every one of us subconsciously retouches the picture of his or her life to make the grotesque travesty of living-to-die make some sense and seem reasonable. We learn to live with a finger in the flame of doubt by refusing to acknowledge the pain to ourselves and each other, or to look for the cause of it. We are benevolent liars to our children, hearty cowards to ourselves. Apart from a few self-realised individuals, only the insane see the unbearable truth. Ours is a jungle in which everyone is kept as busy and unquestioning as is humanly possible, not for common welfare but for common safety. Only in the real jungle there are no policemen, no doubts — and no insanity.

The unbearable truth, which few are allowed to realise without going mad, is that every man is fighting or trying to live through his nature in order to make a better or more meaningful life for man to come. This is the dynamic of human evolution. Each life is lived solely for another life. As the point is always another life there is never really this life — therefore no apparent point, no permanence, no peace for long, and no death. While man lives, he must fight and strive to build the individual he will be tomorrow. As there is no tomorrow, and yet only tomorrow, there is no end. Man can never rest content, even when he is dead. This is the epitome of futility, insanity. Which is why for man who is mortal, the truth of life must always remain myth.

HOW VIOLENCE GREW SOPHISTICATED

Three thousand years ago, man had far fewer interests to keep him occupied. Practically nothing that most western people use, do or think about today was known to him. Most men were serfs, soldiers or slaves. Their wants were simple. Their lives were simple. Their means of killing were simple. Not because their fighting natures were any less acquisitive or aggressive than ours, and certainly not because the fighting spirit was more in control, but simply because in those days there was far less for man to want.

Even in the most distant days, long before the time of serfs, soldiers and slaves, man was neither more nor less happy than he is today. But he was more contented. Contentment exists in those moments when man is not aware of wanting anything. The number of things he can want — the objects of desire which disturb his contentment — depends on what he knows to be available and feasible for him to obtain. This also determines the violence or force of his wanting.

Wanting is violence. All men want. So all men are violent. Man by nature is just as violent today as he ever was and ever will be. But the pressure, the force of his natural violence, becomes more and more diffused, less compulsive, as he discovers more to think about or dream of having. A man who wants numerous things all at the same time is not a violent man. He may be discontented, irritable, restless, industrious, an ordinary citizen. Only when his wanting has concentrated into one intense desire does he become dangerous.

Modern man, with so many desirable things within his reach, is able to diffuse his wanting-power through thinking and imagining. Thought and imagination become sophisticated outlets for the natural force of violence within him. By harmless thinking, planning and day-dreaming he spreads his violent wanting-energy. He diffuses it even further by talking about his many interests to other people with many interests.

Civilisation comes with sophistication — the art of concealing forceful wanting. And as civilisation, sophistication and social coherence become more and more the way of life through the growth of information (things to want or talk about) so the origin of it all, violence, becomes more and more removed as a necessity of life. Torture, flogging, capital punishment and all the associated cruelties become publicly unacceptable and are abolished. We are appalled at the violence that even the last generation tolerated as necessary, let alone the iniquitous times which condoned the Inquisition and all the other horrendous atrocities committed in the name of man's law. Yet that law, so often infamous, constantly moves forward towards the more humane and less violent ends of civilisation. To some future man we will have been the twentieth-century barbarians.

Back in the earliest of days, when man had very little to dream about or imagine as outlets for his wanting, the spring of violence in him was a simple, savage one. In between his few wants he was more contented than modern man but when he did want it was with tremendous willfulness. All the violence in his nature ripped along that one channel. His psychic mechanism was like a steel trap and one touch of wanting would set it off. Early man could not be trusted to remain non-violent for any length of time. He was utterly unpredictable. His hours of contentment would suddenly snap into violent action as a desire arose. He grabbed what he wanted, or fought for it, with the pure savage fury of a man or animal who desires only one thing in that moment.

When man eventually had more to want and became more imaginative, he was less erratic. The unpredictable savagery was on the wane, but his violent nature remained and the force of his wanting was still very much the same. In the later days of serfs, soldiers, and slaves, man's rulers understood his simple psychology and exploited it. By keeping him underprivileged, as a slave for instance, man could be made to give up wanting the impossible things possessed by his masters, and be content only to dream about them. A day-dreaming, imaginative man is not a violent man. He is no risk to established privilege, as long as he can be prevented from wanting.

The birth of shame

The man who wants the impossible is a dangerous man. For the 'impossible' is only what some other man has. And if that impossible thing happens to be freedom, and if freedom is the only desire a man has, he is indeed a fellow or force to be reckoned with. Otherwise the most powerful want a man can have is for the thing he sees as the symbol of his freedom. The most ancient invading armies were raised on this psychology. The soldier was not fighting for his freedom: he imagined he already had it. The soldier demanded a more tangible symbol of freedom: spoils — the right to indiscriminate pillage, rape, wine, gluttony and all the rest of it, including the fleeting posture of absolute power (the caprice of life and death) over his defenceless equals, the conquered army. He wanted the power over others that his masters had over him without the fear of penalty or the need of justice.

By inciting and encouraging the soldier's imagination with these orgiastic symbols of freedom, his privileged rulers kept his primitive emotions whirling with anticipation. Spoils were one thing man could want with all his unsophisticated heart and soul. These substitutes for his usual contentment could be depended on to drive him on through all of war's privations and hardships to the moment of victory or death. His masters indeed knew how to harness his wanting-power, the fighting nature: promise man any freedom he can be persuaded to believe in and he will do anything for it.

But eventually the soldier's fleeting spoils and wretched keep were not enough for him. His discontent, like his imagination, was more lingering. He began demanding more. His sophisticated, privileged rulers quickly got the message. Before the soldier could even think about organising the mass wanting-power he had unknowingly discovered (revolt) his masters smartly stepped in and invented money for him.

It was a masterpiece of evolutionary strategy. Man took the bait — incapable of snatching the rod and the power behind it. With money and promises his rulers began playing him and paying him off. Not for many, many thousands of years would he ever again think of, or want, the impossible freedom which he came so close to discovering — power without money.

The appearance of money in the world signified man's social realisation of shame. It was a momentous point in man's evolution. Until money was invented for his use, he had not wanted enough things to be ashamed of what he wanted. His needs had been too natural and basic for there to be much choice. What one man wanted was pretty much what all men needed; and what all men want is nothing to be ashamed of. Shame arises when some men want what most other men do not choose to be seen wanting. Money's base appeal is that it does not reveal where it came from, or what kind of act earned it. Thirty pieces of silver could be payment for any job. Spoils, like other men's possessions, and even power, can be too easily traced back to a source — perhaps back to a donor grateful for some treacherous or infamous act performed for him by another. But money, that superb alternative to possessions and power, can be hidden away or just spent. It does not identify the giver, the receiver, or the act that earned it. Money is universally faceless. It represents in one idea everything a man can desire or want — a very dangerous commodity.

So now the soldier fought for spoils plus money. His dirty deeds — becoming more and more numerous as he found more things to want — could (with a little luck) be concealed. Men did things to satisfy their wanting, or paid for them to be done, and pretended they did not do them or want them done.

Out of this pretence and the shameful feelings it evoked about his wanting, a new and pernicious kind of dishonesty crept into man's mind and his world. For honest toil man was pleased to accept as barter his keep or staple commodities. But for the killing and exploitation of his fellow man he needed a traceless hard-bitten currency — money.

Money is looting. That is why, since they invented it, man's privileged rulers have never touched the stuff.

THE ART OF CIVILISATION

Natural selection insists on the survival of the fittest to perpetuate the species. It is the law of the jungle and it works — in any kind of jungle, natural or concrete, where the killer instinct is necessary for self or pack survival. Civilisation is not concerned with self or survival. It leaves this to the natural acquisitive and competitive instincts which look after themselves. Civilisation is a cosmic pressure, not an instinctive drive. It does not affect animals. Animals and the least evolved men are content with habitations.

The civilising process at all times represents the descent of higher mind into matter, of human spirit into instinctive nature. This descent brings with it the ideas responsible for material progress. But the nucleus of these ideas, the principle supporting them, is the ideal of Man. It is against this ideal that the achievement of every civilisation is measured: to what degree did it succeed in providing justly for *all* its people?

Civilisation is art on the grandest scale, the continuous endeavour to express excellence in the form or matter of humanity. Humanity is Man's becoming.

Civilisation is the art of putting man first. This is not easy to comprehend let alone to try to put into practice. It is not just the admirable concept of putting the other fellow first. It is not just being kind or doing good. It is best described as doing nothing more than being just. Civilisation demands that all human beings within its compass must come first — or none at all. All the people are the one Man. However, at this stage of evolution, human beings are only part Man, as society is only partly civilised.

Man, the idea or character of the human species, is a towering, impeccable principle, fully alive in all human matter and human circumstance and beyond any one person's intellectual or self-conscious capacity to know in its entirety. Yet each human being can be that Man, and each society can be that civilised excellence in any moment of nobility, justice or compassion.

The nearest the individual (or a society) can get to understanding the principle of Man is through that person's own ethic or ideal — how each one of us would like the world to be for ourselves as our fellow-man. Service to this ideal through the means at our personal disposal (and as a community, at our collective disposal) is the civilising process.

DECLINE AND FALL

All civilisations are attempts at civilisation. All have failed due to an inherent flaw in the social conscience stemming from humanity's inability to grasp what civilisation is for.

Within the more impressive achievements of each great civilisation is the means of its inevitable destruction. Where it was thought to have succeeded most is where the poison will be found. It is like a parasite that gets into the flower in full bloom and infects the fruit before it has formed. The Romans' genius for colonisation and government was the means of their downfall. For the earlier Mediterranean peoples it was rampant curiosity about the external world; seeking distraction, they gradually forgot man's astonishing origins which were the foundation of their culture. They subverted the power of that knowledge — original art — and created beautiful ideas and things instead of excellent men. This ignorance and moral laxity finally allowed the barbarians to invade their territory, as well as their psyche, and eventually destroyed them. Vandalism in all ages lives on in the hills of the mind.

Many other civilisations, such as the Inca and Aztec cultures of South America, collapsed because the people were indoctrinated by priests who turned their myths into superstition. Rule by priests prevents the necessary development of discrimination and common sense in worldly affairs which is part of survival and the struggle for civilised excellence.

The destruction of our civilisation, western civilisation, will come through its super-technology, probably in some form of almost instant wipe-out as a logical improvement on the older, slower process of degeneration and decay.

The means of destruction of a civilisation must not be confused with the cause of destruction. Moral failure has been the cause of every civilisation's downfall. Our moral failure, the worm in the western flower, was cultured by the drive among the privileged to have more while the many by comparison had little or nothing.

Ours is the first world civilisation. In that we are unique and represent the end of a phase in man's development. Western civilisation is an extension or culmination of all earlier civilisations and their aspirations for ever-more conquest, wealth and influence. But now western civilisation has gone to the limit and conquered the globe. The conquest has yet to be consolidated and consciously realised, if there is time. But the pattern of consolidation is inevitably established and there can be no major changes, just modifications. Any pockets remaining of the ancient cultures are already regarded as backward or underprivileged; and by their own consent and connivance are being eroded further every day.

Our main civilising achievement has been to bring humanity together through our technological and intellectual genius without anybody needing to change their position on the globe. Through western thought, politics, weapons, transport, finance, science and telecommunications we have created a westernised world civilisation, exciting everyman's wanting and ensnaring him with his own expectations. Each day the world gathers more solidly under the flag of intellectual materialism, in spite of its ideological differences. Never has there been such solidarity. The common aim is for more, more of everything, more freedom, leisure, government, power, peace, military hardware, information, travel, flush toilets, drugs, computers and unlimited credit. There is no other way to go and no evidence of any meaningful demand for an alternative. The remaining obstacles to the completion of the westernisation of humanity are not due to any outraged opposition but to sheer lack of the resources to satisfy the universal scramble, whether white, black, brown or yellow. The world is populated by westerners: first, second, third, fourth, fifth, sixth and seventh-class citizens; in other words man's privileged rulers, the super-rich or powerful, the wealthy, the workers, the poverty-liners, the exploited and the abandoned. These are the schisms of failure in western civilisation.

Ours, like all the great civilisations before it, has served the planet's evolutionary cause. As a part of the cosmic earth experiment it has succeeded. But its failure, and unfitness to endure as the permanently civilising way of life, lies in the expediency of its shifting values. Our particular failure is intellectual duplicity, our double standards. In two thousand years western civilisation has not been able to be honest to itself, to its people. The times we lived in would not permit it. By the 'times' I mean the power mankind has attained in any age to control matter, or circumstance, morally. Man who can control matter morally is a man who puts man first.

We have not possessed the moral strength to stand against the pressure of wealth and self-interest, to uphold in practice the worthy ideals we protested. Where men rose to rule, the ruled became the sacrifice. Now, after two millennia of opportunity, no man or men can ever again rule in this civilisation. The 'rulers' are ruled by circumstance, the force in matter. They are powerless to change anything significantly, as previously was possible — and the thoughtful ones know it. The privileged position of rulership on this planet in this civilisation has been forfeited and withdrawn.

A COSMIC LAW

Man, in spite of his nuclear might and biological villainy, still does not possess the power to wipe out the human race. But he can destroy most of it. And, as I have said, this will be necessary for his evolutionary advancement.

Man has a limited mandate for violence. Although he cannot destroy the race, he can decimate its component parts, as he has so often demonstrated in the past. The scale of violence man can inflict on the planet or against humanity at any time is controlled by a cosmic or deeply subconscious principle. This principle also operates against any internecine accidents or folly which could threaten the existence of humanity. Such a profoundly important cosmic issue as the survival of the human race could never be permitted to depend upon an emotionally capricious will, such as man's at its present stage of development.

The cosmic law now guaranteeing the survival of the race against man's destructive genius and fighting nature states that his ability to destroy nature or himself (as humanity) at any time is only equal to his knowledge of nature or himself. In other words, he cannot destroy himself completely until he knows himself; and then he will know better.

The demonstration of this cosmic law is in man's weapons. They are his evolutionary status symbols and reveal the extent of his knowledge of himself or nature. Once his ultimate weapons were the club, knife, axe and spear. They were primitive like his knowledge and aspirations. He understood very little about nature or himself. And so, in line with the cosmic law, his ability to destroy either on any large scale was correspondingly limited. As his knowledge, ambitions, imagination and skills increased, he added the sword to the arsenal, enlarging his lethal capabilities. Then came the long-bow with which he could stand away from his enemy and destroy greater numbers with greater impunity. All the time his knowledge and power to want (to visualise more) were increasing. Eventually, delving with his mind into chemistry, he discovered the knowledge of gunpowder, the first significant means of mass destruction. This matched the revival of learning and artistic aspiration represented by the Renaissance. Over the next few centuries, as he probed deeper into nature's secrets, becoming more aware of the finer things he could make and want, man developed ever more effective weapons and systems of mass-slaughter — artillery, the cartridge gun, repeating rifle, machine-gun, gas, aerial bombing, rockets, guided missiles. Then through atomic physics his knowledge of nature was sufficient to devise the Bomb. Today this appalling weapon of mass destruction, or mass knowledge, can devastate (but not annihilate) nature and humanity — a fitting symbol of man's enormous knowledge, and matching power to inflict violence.

But still, in spite of our fears, the ultimate weapon has not yet been developed. Man's knowledge of nature or himself is not yet up to it. The ultimate weapon, symbol of the ultimate knowledge he has yet to achieve, will be to use the earth itself as a bomb or energy device and to disintegrate the whole planet. One day, many scores of thousands of years hence, this may be done deliberately by cosmic man.

Just as cosmic law controls the devastation man can wreak upon the earth at any time, so cosmic restraint has so far prevented him from unleashing the nuclear force now at his disposal. The balance is so fine that nuclear devastion is a possibility at any moment even though it subsides periodically as a perceived threat.

The only certainty — guaranteed by man's current state of self-knowledge — is that he will survive in sufficient numbers to perpetuate the race. This is hardly a comforting thought for the billions who will be destroyed; nor for them to know that the holocaust will have been absolutely necessary for the inevitable progress of westernised culture towards the enlightened race man must become.

The cosmic law I mentioned only guarantees the survival of the race; it cannot determine whether or not man will use the Bomb. Determining its use is the distinction between power and force — the two foci of the cosmic moral system in which man is involved. Down the ages, from knife to Bomb, a fine moral element has determined whether man has used the ultimate weapon of his times. This involuntary moral control lies in the quality of man's wanting.

The more man knows about life or the world, the more there is of it to want as power and possessions. Yet the more a man lives or practises his wanting — which can be a violent time for himself and everyone else — the less he will eventually want; and the more will he be 'civilised' or restrained in his desires. We see this not only in our own lives and in those of the people around us, but also in the struggles of nations to become developed and prosperous. A nation with enough knowledge to have made the Bomb has a great deal of wanting-power. Today, with many countries sharing the knowledge or wanting-power implicit in the force of the Bomb, the need for self-control or civilised restraint (victory over the fighting nature) has become imperative.

Here is where the the moral dimension of the cosmic law comes in. So far man who controls the Bomb has been civilised enough not to want anything the Bomb can give him except the peace of being spared its immediate and cumulative horrors. Unlike every other weapon in history, even for man's privileged rulers, the spoils have not been worth the consequences. The significance of power is now revealed. At this level of non-action — negative wanting — the Bomb becomes a power for peace instead of a force of destruction. This power, the product of restraint, becomes the power of non-violence referred to earlier, like that of the true master of martial arts who has the power to destroy but can never use it. The release of power as force at any level of existence invites counter-force, counter-blow or counter-tactic — the misery of war and conflict.

But one essential weakness remains in the peace-keeping power of the Bomb. In depending on it for his survival, man is relying on a power outside himself. The Bomb symbolises faith in the power of the weapon, the western faith now fully materialised. No one individual has control of the weapon in the way a man can be said to have control of himself. The situation is therefore precarious and through nuclear proliferation becomes more so every day. It needs only one power-mad individual or nation to demonstrate the horrific consequences of the western philosophy of developing the weapon before the man. Sooner or later some idiot, or group, not wanting the peace the Bomb offers, is almost sure to push the button.

The most powerful man is he who knows a lot but wants nothing. He can protect us. The most dangerous man is he who knows a lot and wants a lot. He can blow us up with no need for it. A man who knows a little and wants a lot is a novice or a fool; and just the kind of idiot who will be used to do the job when the time comes.

PARANATIONAL TERROR

The newest and most world-shaking extremity of violence and change is what I call paranationalism — any organised attempt through terrorist acts to draw attention to the evils of the status quo anywhere in the world. It aims to destroy the root-attitudes of people supporting the formation of power and privilege —

conventions so integral and fundamental to a society that even to the radical political mind the attempts seem senseless and even psychotic. This is especially so since the violent means employed usually involve the death and injury of innocent people.

Pure paranationalism does not preach revolution, anarchy, nihilism or ideological fantasy. It is not idealistic in the least. Using indiscriminate violence against the expectations, values and unquestioned loyalties of western society, its strategic weapon is violence itself. Instead of taking on the police, paranationalists take on the politicians and society's leaders as a means of getting at the people. They know that no one can get directly to the people, who in a modern society are completely shielded from change by the mass media, the authorities and social convention.

For the paranationalist the people are as much the enemy as the objective. This seems a contradiction, but to the dedicated terrorist there are no people in the western world, only positions. In his mind, people act out their positions. Positions have to be destroyed in people's minds for the real person, the individual, to be reached. So the terrorists show no mercy, no compassion of the kind that conventional people in conventional positions relate to.

The paranational terrorists are a new breed of suicide fighters, like the kamikaze warriors of Japan (whose name translated means 'divine wind'). To the paranational terrorists death is nothing, and living is ignoble once the evil of positions-without-people is seen.

The terrorists' motivations cannot be comprehended by the conventional mind or western attitudes because they operate at a subconscious level; they represent a new psychological phenomenon rising out of the unconscious in man. Its only purpose is to destroy the certainty of the westernised mind.

Paranationalism depends on the stark terror and horror of its indiscriminate actions to drive home its idea or message like a steel pin into the subconscious of humanity, through the armourplate of the enemy — the contented mass position. It aims at the individual, not the masses. Every terrorist act has a subliminal message that lodges somewhere in the human psyche where it will slowly rise to consciousness among the younger generation, tomorrow's rulers. But from the western viewpoint, the terrorists' motivations, like their achievements, are inscrutable.

There is no conventional power-drive in this kind of terrorism, no personal reward apart from death. As the precursors of an approaching new culture, paranational terrorists tend to come into western civilisation from the geographical East. Where the westernisation process is most advanced and the social conscience is most likely to be outraged is where terrorism will strike most hideously and most often. But paranational terror is not restricted to democracies. As a new expression of power it is currently mixed up with nationalistic and religious struggles on every continent. It has yet to divorce itself from these conventional freedom struggles and realise its independent identity and task — as it has yet to do its inconceivable worst.

THE BEGINNING OF THE END

The first shock-waves of paranationalism, unnoticed or forgotten by the majority almost as quickly as they disappear from the headlines, represent the beginning of the end of western civilisation. The millennium of decline is here. It is difficult to imagine that our familiar way of life can possibly come to an end as a way of progress. The populations of past great civilisations probably felt the same. No civilisation seems able to visualise its successor. And yet all lie in dust apart from a few relics.

The time that is left is virtually no time at all. Just before the end there may be a final, brief cultural flowering, the natural result of our civilising achievements. This flowering will be distinguishable by the feeling in the individual or the society that life has reached some worthwhile peak.

One day it may be said that western civilisation gave more men more years to live in better health, comfort and luxury — so that finally it could destroy more than if it had never existed. Whatever the means of destruction, through terrorism, nuclear devastation or some other cataclysm, the barriers that separate the different nations and peoples will fall down. They are intolerable anachronisms in a world civilisation.

Evolution proceeds on this planet haltingly, in distinct stages of time in which the process stops, gathers, bursts . . . and out of the shambles and chaos leaps the Zeitgeist, the spirit of the era, to seed a new attempt at civilisation containing that which was best of the old.

Five

THE
UNDERWORLD

*What is the origin of fear, dread
and all emotions? In mastering
them man undertakes the next
great evolutionary task*

THE UNDERWORLD

THE GENIE GETS OUT OF THE BOTTLE

MAN'S MOST CIVILISING drive is the need for pleasure. His most self-destructive impulse is the need for love. Between these two compulsive longings lies the treacherous ground of emotion, which he often mistakes for both.

Let me make it clear what I mean by 'emotion'. Man is being emotional, and therefore not himself, when he chats or thinks about the past; when he tells his troubles to anyone except perhaps a professional consultant or spiritual teacher (both of whom will keep him to the facts); when he gossips about another person; or when he justifies himself or accuses or blames another. He is even in danger of being used by emotion when he expects anything from another; or when he makes a statement not in answer to a question direct or implied. When I speak of emotion I am not just talking about the stronger feelings of anger, jealousy, grief etc. I am referring to something more subtle, yet massively pervasive — something as fundamental and ubiquitous as the past itself. In fact, emotion is the past itself.

Emotion is the past because it originates from instinct, the first driving-force of life on earth. Instinct is the desire to survive and live again. Its very essence is the pure knowledge of having lived before. The finest accomplishment of instinct is seen in the life of the animals — so complete and perfect in its unselfconscious artistry. In man instinct continues to govern his animal responses but when it rises into his unique selfconsciousness it becomes emotion — the need to project himself and make an impression on his environment, whether by anger, mood, talk, threat or desire for recognition, sympathy and acceptance. Emotion is the self-conscious expression of instinct. Animals, being unselfconscious, do not get emotional.

Instinct was the driving force behind the development of the animal body to the point where selfconsciousness was a latent possibility. Similarly, emotion was the rising wave on which man travelled out of the psyche — until he finally became attached to his physical senses and to the world as his awareness of self. Then emotion stimulated his mental development to its present creative level of performance — mainly by forcing him, through his competing desires, to think and choose.

Now that man has realised his senses, entered his body and ordered his reflective thought processes, emotion as the prime mover of these evolutionary achievements has served its purpose. But like the legendary genie who escaped from the bottle and discovered its enormous power over man, emotion cunningly refuses to go back into the bottle. It is now man's torment, the main source of his misery and lack of fulfilment. Emotion is living off him, sucking like a vampire at any excitement it can stir in his sensitivity; leaving him depressed, discontented or unfulfilled, until he is revitalised and can be used again.

Nevertheless, as it was evolution which extracted the cork and let the genie escape, so evolution must redress the position. How to get the emotional genie back into the bottle, and put the stopper on for good, is now the great evolutionary task of humanity. Consequently, more of the human race than in any other time are searching for solutions within themselves —

without realising the identity, subtlety, power and cunning of the one common enemy.

Modern man is endeavouring to rid himself of the confusion and uncertainty created by emotional indulgence and ignorance. The remarkably popular self-growth movement, all the different meditation methods, therapies, schools of philosophy and psychology are directed towards this end. A huge and growing section of humanity is ready to discover that for all its apparent sincerity and harmlessness, emotional living and thinking is no more than pathetic, self-indulgent games-playing — the demand for love. Love cannot be demanded. But emotion thinks it can.

By facing up to the fact of emotion, by starving the demonic genie out of himself, man can begin to get free of its selfish use of him. Then he can begin to realise and fulfil his two basic drives — for pleasure as self-expression, and for love as union. As he is now, under the influence of the distorting and confusing demands of emotion, man can only vaguely sense that continuous creative pleasure and constant love are attainable. Certainly, his experience of living does little to confirm this; creative pleasure and constant love remain for most an impossible dream. Yet everyone in their own way continues to pursue them with unshakable certainty. For the simple truth is that man was born to love and to create.

INVASION FROM THE DARK

Emotion was created by the action of sunlight entering and exciting the unconscious psychic mass in primitive man. Being the first child of light, conceived in the black womb of the unconscious, emotion's prime motivation is to get back up to the external world of sunlight, back to the source of its excitation. It does this by rising and taking over man's brain or senses. But its natural place is in the inner dark. It has no place in the outer world; here, out of place, it creates conflict and chaos.

The demonic genie of emotion is restless and immature, unseasoned by sufficient experience of the outer world it seeks. It continuously rises to the surface like an inquisitive snake out of a dark underworld. Given the chance, it peers through the unsuspecting eyes of the human being into the selfconscious outer world. And it looks for recognition in the mirror of the world, for an emotional reaction, its own existence affirmed in someone's eyes, face or attitude. That is why it is dangerous to look into the eyes of some dogs unless you intend to deal with them. Our selfconsciousness makes the animal's otherwise purely instinctive reaction, emotional. The only protection is to remain perfectly emotionless in your eyes — a very difficult thing to do. If the animal senses the slightest weakness or fear in your eyes (the presence of selfconsciousness), and it feels it can win, it may attack.

When we catch the eye of a person who is emotional we instinctively know that they are looking for trouble and we should stay clear. Emotion is trouble looking for itself. Emotion is a violent energy which cannot resist the opportunity to enslave, hurt or destroy anything that is emotionally weaker. It is cannibalistic. It lives off itself.

Emotion is fairly easy to discern in others. In ourselves it is much more difficult because of the speed with which it takes us over. Before we can be aware of it, emotion is there, performing in our own awareness as 'us'. If we can manage to race ahead of it, or more realistically, develop the power in ourselves to be intelligently present the moment the genie begins to rise, we can halt it and master it. This can be done without the frustrations and neuroses which come from bottling up the genie in ignorance.

HOW THE SELF GETS INTO THE BRAIN

In primitive man emotion carried the waves of instinctive intelligence up and out to the unrealised senses. It pushed and it strained ceaselessly from within to bring the sense of knowing, the sense of self, into the incipient human brain and then into the outer senses. When the last connection was made linking each sense organ in the outer world to the physical brain, and to the psychic double within, man not only possessed the instinctive senses but he knew he did. As soon as he was able to reflect that he could see, hear, smell or touch something, his sense of self was born.

Alone of all the species, man now had the incipient ability to think reflectively — to be developed as a survival tool. He had to think positionally to start seeing his self, his body and emotion, at the centre of things and to make mental connections with the rest of the environment from there. Previously this had all happened instinctively with no awareness of himself.

The positional mind was the psychological extension of animal instinct. It paralleled the evolution of the human memory. The early positional mind lacked the subtlety that came with self-consciousness and the inflow of emotion. Primitive man's memory, had none of the fluidity in making connections that our mind has today. Primitive man 'thought' disjointedly. Each memory he had was separate, solid, very concrete. He recalled only isolated past incidents, as fixed perceptions or positions. For example, if he caught an animal for food or found water or any other need in a certain place, he would remember that event in isolation and return there habitually, even though what he needed never turned up there again. This early positional thinking established wandering and nomadic patterns which whole tribes followed without ever understanding their origin. The patterns were later given magical and religious significance leading to ritual; and those who remembered them the longest, the elders, were named wise and given venerable titles.

With the advent of selfconsciousness, man used his positional mind and fixed perceptions in an attempt to interpret and satisfy new emotional feelings or desires as they arose in him. He enjoyed letting past events play on his memory and watching the pictures come up. Gradually he resorted to the past in this way to deal with the present and started the positional habit of living in the past which persists in all men and women today. This was the origin of the old, old habit of 'naming' which binds us — and those we name — to the past.

Naming is judging. It is a pernicious habit of the self and operates incessantly in man at the conscious and subconscious levels. It is always negative in the sense that it replenishes the emotion or past in the self and reaffirms its existence. Naming or judging is so subtle that it even extends to aimless thinking and worrying: you cannot think or worry without naming or judging.

Naming begins by man naming or judging himself. He thinks, 'I'm a failure', or 'I'm not likeable', or 'I'm unkind or guilty'. Or he names or judges himself to be good, handsome, superior or a jolly fine fellow. These judgments are still negative as they inflate the emotional self and help to keep it judging and naming. Further, the self is constantly judging the immediate environment. It subconsciously likes or dislikes what it is seeing: that's good, that's not good and so on. To break the dependence on this primitive, positional mind and cease the constant naming involves a good deal of conscious effort. But when a man or woman is able to do it, that emotional part of the self dies.

As self is created out of emotion, it is all past — even now. The 'now' that the self recognises as now is really a spurious 'past now' — because self relies on memory and naming. Along with the death of self goes that old 'now', which was reference to the past or living in the past and the main source of confusion and uncertainty.

When the naming stops, man breaks out of his narrow time-bound cage. There is a refreshing new sense of time — the moment, the real now. This moment is minutely ahead in time of that old self with its spurious 'now'; personal time has in effect moved 'forward'. Now he is truly alive and no longer just living. Now he sees things and people as they are, not as they were.

This 'moment' is part of a new factor of awareness currently making its appearance in consciousness. It is extremely fine and delicate as yet; perception of it is clouded over and lost immediately there is any emotion or subjective reference to the past. The moment is based in intelligence, or the purely mental, as opposed to the old self's emotional base. To exist as a ceaselessly naming individual is emotional and unintelligent, therefore expendable.

In any moment, with conscious effort, we can observe the naming habit in ourselves. But remember, it begins by naming or judging yourself; that then leads to the judgment of other people and other things.

THE VITALITY OF THE PSYCHE

The human psyche is our instrument of thought and perception. It begins immediately inside the human body and extends down through the subconscious (where we think and dream) into the unconscious. To us who are the dying living it is unconscious because we cannot be conscious of all that is in there. This part of the psyche includes the world of the living dead and, deeper down, man's psychic double.

Each living creature, from an individual man to a collective organism such as a colony of insects, is attached to the psyche like hairs to a scalp. In total, the mass of all living things on the earth at any time covers the psyche, or psychic body, like a short dense pile.

Every living thing is a receptor, convertor and battery serving the evolution of the total psyche. While alive, each organism absorbs into its unique psychic system (its 'hair', or body) cosmic energy and data provided by sunlight, the dynamic behind all life and terrestrial evolution. From the energy of light that it has assimilated, each living battery transforms and stores two primal life charges, one positive/actual, the other negative/potential.

The positive charge is the vitality which powers the life of the physical organism. The negative charge is far more subtle and impossible to quantify. It represents the intrinsic value of the life of the particular organism. The value of each bit of life or life-experience, no matter how ephemeral it seems or how minute in form, is equally important and essential to the whole psyche. Throughout the lifetime of the organism the negative charge builds up and remains unused. At death the positive vitality remaining in the living creature/battery travels down the narrow channel of its stem or 'hair' into the collective unconscious of all life; and it takes with it the creature's unique negatively-charged 'essence' — each organism's contribution to the experiential value of the whole; the value of its having lived. Then, in an intensive after-life process the negative essence is extracted and the positive vital energy rejoins the collective reservoir of recirculating instinct. After a further process, the essence takes on a new vital potential and, returning to the earth, manifests in a new body among the species; and the whole cyclic process is repeated.

The most extraordinary fact about the psyche is that it consists of the instinctive and emotional past of all life that has ever lived and died on earth. Every creature since the first micro-organism is an evolving part of it. At every moment the psyche is being swelled but also refined by the continuous recirculating life-and-death cycle of all living things.

THE SOURCE OF DREAD

The psyche being used to absorb and understand these words has evolved over the past two thousand million years out of the instinctive surviving vitality or incipient emotion of all life on earth. This vital consistency of the psyche explains many of man's otherwise mystifying and irrational dreads and fears.

The psyche is one vast fluid whole. Man's psychic double is a centre of consciousness within it, his most intelligently developed nucleus. Depending on how mature he is in evolutionary terms, the double is either conscious or more or less selfconscious. But this relatively enlightened human centre is very small compared with the enormous surrounding mass of natural or instinctive ignorance. Instinct, in spite of its superb efficiency as a survival reaction, is ignorant and unenlightened in relation to the movement towards a civilised or moral purpose which only man among the species consciously reflects. Consequently the individual man, in spite of his 'superior' reflective intelligence, frequently becomes psychically disturbed by the pressure of the surrounding primitive mass. The dark hours of early morning are often the most anxious and depressing for him. These 'dark hours' are no more deficient in light than other night-time hours, but they have the power to disturb, arouse dread and exaggerate fears because they represent the nadir of man's day. When he is denied the power behind his selfconsciousness (the reassuring, restoring stimulation of sunlight) the surrounding psychic ignorance can assert itself and assail man's sensitive centre; he feels the vulnerability of his beleaguered psychic position, usually with pessimism and sometimes with hopelessness.

The sufferings of all men and all life that has ever lived are indelibly imprinted on the psyche. The experience of every horrible death, cruelty, torture and pain persists as a vital part of the whole, intermittently rising enough to be reflected in the selfconsciousness of the individual. The superstitions to which even the most civilised of men are prone have their origins in the dread of the unknown, which really is the dread of what the total psychic consciousness has already known. Most nightmares, psychic disorders and apparent insanity have their source in involuntary perceptions of this red and raw unconscious region of the psyche. In the dark hours the hoot of the owl, the howl of the dog and even the human voice can induce fear and dread, for every cry contains the hidden resonance of countless ages of animal and human suffering. The hallucinations that overcome the schizophrene are mostly due to these perceptions.

PSYCHIC POSSESSION

Use of humans by psychic forces is extremely common. Since the psyche pervades man's brain and every cell of his body, these forces are able to invade the individual's psychic space with imperceptible stealth. They enter from a man's psychic double via any emotional abnormality.

The presence of the psychic forces excites him; and the excitement enables them to take temporary (sometimes permanent) possession. In this way the energies receive the vital stimulation of experience — something that cannot be gained directly in the psychic world — without having the responsibility of existence. When the invading entity has had its fill and the passion or compulsion is exhausted, it vacates; leaving the person to face the consequences, usually in a devitalised state. A feeling of being cold or drained often follows.

Just about every negative mood or feeling we experience can be attributed to this type of possession. One strategem of the energies is to induce and then attach themselves to thinking. Thoughts originate as single-frame flash images which are unavoidable. The danger arises when a single thought is allowed to run on by association — thinking. Invariably the thoughts focus on an emotional point in the past. The psychic forces wait on these thoughts, enter on the emotion and temporarily influence the moods, decisions and actions.

The psychic energies also attach themselves to man's fear of illness or death. When he is sick they make conditions worse by causing depression. They can attach themselves to natural and involuntary processes in the body, such as a woman's period or the effects of ageing. In this way they can manifest as brooding or hysteria, or even paranoia.

Over-indulgence in sex and other excessive sensual or emotional stimulation can provide the necessary catalyst for psychic forces. Any over-excitement in anticipation of even quite normal events (as often happens in children) can lead to temporary possession.

The after-effects of possession are depression, self-pity, tearfulness, pessimism; being moody, demanding, or lethargic. These effects may not appear until a couple of days later, which shows the cunning of the psychic entities. The delay prevents the person from seeing that his depression is caused by his earlier thinking or excitement. If the person realised this it would undermine the entities' parasitical existence. What I am saying is of course incredible. Everyone knows that thinking is natural and perfectly harmless. And that's the lie.

Protection against psychic invasion lies in resisting any thought that does not involve the intention of action. When action is the object, such as in preparing a plan and performing a task, man reasons from fact to fact and is not emotionally vulnerable. Any other thinking is discursive and carries the danger of psychic invasion.

THE RETURN TO IMMORTALITY

The point of existence is for man to build a new world of his own and be fully and permanently responsible for it. It has to be a world in which he is immortal, where his conscious presence and control are uninterrupted. Men of goodwill have sensed this down the ages. Many have imagined it as a hope for the physical world, only to have their efforts ended by death or uncontrollable circumstance. Mystics and saints have seen it as a life in the spirit of selfless love, in which the need for any existence at all disappears. But man's future world is neither physical nor spiritual; it is psychic.

While man is alive, he must start to take conscious possession of the psychic world, currently the mysterious and turbulent domain of the dead. To date, he has had no world to call his own for very long. Death excludes him from the physical world; living excludes him from the psychic world. He never remains long enough in either, is never sufficiently responsible for his collective actions to be entitled to call one or the other 'his'.

The psychic world as a whole is chaotic. It is man's own involuntary creation, in some ways his monster. Together with his accumulated feelings of love, devotion, kindness and goodwill, it includes all his pain, greed, cruelty, hatred, jealousy and sadism. All these emotions are vitally alive there and able to take form by entering any of the countless mental images, from the saintly to the bizarre, which the primitive and developing mind has put there. Nothing in the past has been lost in the psyche. Man must take responsibility for the whole. All must finally be resolved by and in man, its creator.

Man's power to control his destiny in the death world is no greater than what he manages to achieve while alive. While alive he cannot see that the final responsibility for his life and the world is his. He mainly lives a life of abdicated responsibility. He compromises endlessly, goes with the stream and contributes only what is necessary to the common good. Consequently the expansion of man's awareness is glacially slow and his power to change things in either world is practically nil.

The epoch has now arrived in which man is to start taking conscious control of the psychic world and his own existence. This will be done by the individual deepening his consciousness to include knowledge of his psychic double in the world of the living dead. Only the clarified consciousness of the individual has the power to unite both worlds — the worlds of the dying living and the living dead — to form a new immortal existence. Success by a few will make a stronger bridge for others to follow.

A man's psychic double — his own vital and immortal self — is as close to him as his sense of being alive. But he is separated from it by his negative feelings, moods, fears and desires. These negative feelings distort his psychic space and separate him from his immortality. The space becomes the playground for the psychic forces that enter and take possession of it.

As we have seen when a man embarks on clearing his space in earnest, the forces start opposing, afflicting and tormenting him. They are tricky and resourceful, skilled at exploiting emotional weakness, particularly fears and doubts and human ignorance of death. In this contest he can meet the self-willed, perverse, and part-conscious energies — part-animal, part-human, part-divine, part-demonic — that since time began have pressed for self-expression through man and his imagination.

Imagination is the battleground. The further a man gets into the psychic world and closer to his own double, the more imagination becomes psychically substantial. Fears can manifest as psychic visions, particularly at night. They use his own unresolved or unfaced emotions. Consequently, they are as strong as he is weak, as weak as he is strong. Panic is their ultimate strength.

A man ceases to be controlled by his imagination when he stops thinking about emotive areas such as sex, love, power and death. He is then facing up to death and the psychic world. As he does this, the negative emotional forces are rendered impotent and can no longer afflict him; his space is clear.

Many intellectuals feel they have their imagination firmly under control. They are untroubled by psychic forces and are dismissive of this part of reality. But they are protected only by default. They have not yet begun the evolutionary phase of having to face up to death and the psychic reality. When they die they will enter the death world as dreaming shades, not as conscious individuals. From such a mental distance as theirs the reality of the death world appears as fanciful imagery. Such images can be rationalised away, but only so long as the person is able to keep his distance from death, which no one can do for long.

If a man endeavouring to clear his space is deterred by the psychic forces that oppose and resist him, he will certainly be unable to face up to the psychic presence of his own past. This is the guardian of the threshold which he must face before he can be aligned with the unfolding purity of his psychic double. The meeting will without doubt petrify him; if it appears psychically it will probably happen at night. The guardian is a living entity, immeasurably more powerful and terrifying in its perceived presence than a man can ever realise until he actually comes face to face with it. This he must eventually do to gain conscious control of his part of the death world.

Mediums and clairvoyants may communicate with the world of the dead. But no one may have conscious control in it while alive until he has confronted and dealt with the gigantic psychic presence of his own living past at the door to life and reality.

Six

THE
UNIVERSAL MIND

*Now we go back to the beginning, to
see what structures of mind are
at work in creation and how
existence is formed*

THE UNIVERSAL MIND

THE ORIGIN OF THE EARTH-IDEA

THERE IS ONE universal mind. In it are innumerable creation-points
— stars — each of which is the centre of its own stellar mind.

The earth originated in the solar mind as a creative thought
of the sun. This occurred long before the planet appeared in
outer space. For the sun's thought of the earth to become
effective it first had to achieve the status or potential of an idea.
This depended on it being favourably considered by the cosmic
will. Only if it were affirmed by the will would it possess the
necessary potential to continue to mature in the solar mind as a
cosmic possibility in time.

Nothing in the creation process is certain. During the three
phases, from thought to idea to cosmic existence, there is no
certainty that a thought will become an idea or when an idea will
manifest. The most profound thought can be still-born, even
though deriving from a mighty creative power like the sun. The
universal mind is filled with myriads of non-existent possibilities,
thoughts that have emerged from the creative stratum of mind

(the stars) but have not yet been validated by the will. Time and existence are the uncertainty of their becoming or not becoming.

The will is extra-cosmic. That means it is outside the system of the universal mind — the universe, including the earth as we perceive it. Will stands behind mind, entering and pervading it at will. Therefore no mind can know what will is. It is beyond the grasp of intelligence, for intelligence is the condition of mind. (The demonstration that intelligence is the condition of mind is that it varies in degree according to the mind expressing it.) The closest word to describe will is purpose.

Creation involves incalculable time and complexity of action by will and intelligence. The vast intelligence of the solar mind was behind the thought/idea of the earth. But at no level does intelligence know what the purpose of creation is. The creative power is only aware of the idea, the object of its willing. And idea is merely a part of purpose as is any object or objective. Purpose is the will's inviolable secret.

The will's consideration of the sun's earth-thought was not dissimilar to what occurs in the human mind or psyche. When a creative thought forms in the psyche the will either attaches itself to it, or it does not. If not, nothing happens. Some of man's 'best ideas' never mature. The sun's earth-thought did mature. The will embraced it and it became an idea deemed within the purpose of things to be feasible in time. This act of validation by the will allowed the earth-idea to structure and occupy a permanent place of its own in the solar mind.

TERRESTRIAL MIND

The earth is a cosmic being whose mind, the terrestrial mind, is represented by the earth's magnetic and gravitational field. This enormous electro-magnetic field extends out well beyond the orbit of the moon. Its outer region is super-ionised, which means it is so charged and polarised by the forces acting on it (the rays from the sun and other stars) that it is kept abstract. There is no knowing in it, only pure knowledge or consciousness. This mind is unimaginably cosmic and touches on the solar mind. It contains inconceivable future earth existences or cultures in which man has yet to participate and which represent his future potential.

Nearer the earth but still outside the orbit of the moon, the earth mind, although remaining relatively cosmic, becomes less abstract and forms a thin sphere. This consists of a clustered overlay of innumerable disc-shaped concentrations of energy. These are earth-ideas — ideas of the Earth Being itself, permanently fixed in the terrestrial mind.

In spite of the proximity of earth-ideas to each other, they remain pristinely separate in the mind. None influences or blends with any other. Plato glimpsed the sphere of earth-ideas when he wrote of an originating world of ideas behind the physical world.

All objects on earth are the manifestation of earth-ideas. The notion of earth-ideas may be difficult to grasp because our brains (on the surface of the psyche) are dependent on our senses for the identification of objects. We identify objects by their differences and peculiarities. But pure idea has none of these. Car is car, table is table and horse is horse. Colour, shape or size don't come into it. At the level of idea car remains car with no ground for differentiation.

The mind's sphere of earth-ideas is the matrix behind the inevitability of shapes which all classes of things must take in existence. This guarantees their replication with the same astonishing exactitude.

Physical manifestation of an idea depends on the stepping down of the cosmic power of the idea into terrestrial force. This is done through the medium of the psyche, which is a relatively less abstract plasma extending out to the orbit of the moon. Cosmic knowledge or power cannot be transmitted to the earth without passing through the psyche. The cosmic power behind idea is spiritual (cosmic and spiritual power are one and the same) and without the psyche to differentiate it, the knowledge

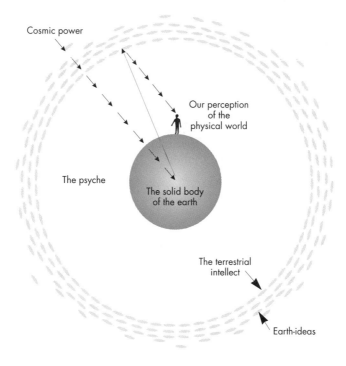

Model of intellect, psyche, world

of spiritual energy would be beyond man. The psyche is superbly sensitive and responsive as a conductor of cosmic power. But the quality of psychic plasma is such that it necessarily refracts and differentiates whatever passes through it.

Immediately on entering the plasma, the power (idea) is broken down into force (information). Ideas reduced to information lose their power to remain pristinely separate. Different ideas become superimposed, one upon another. The idea of horse is modified in the psyche by the idea of brown and other ideas. And the individual horse appears with its distinguishing peculiarities.

THE INTELLECT

The intellect we use is not individual. Nor is it the exclusive property of man. All the species use the same intellect with varying degrees of intelligence. The intellect has a definite fixed position. It consists of the inside surface of the sphere of earth-ideas that surrounds the planet and the psyche. While the ideas furthest out in the sphere are few and loosely spaced, nearer to the earth they are so tightly packed that if they could be seen from here they would appear to form a smooth, highly reflective surface. The intellect reflects all psychic and mental activity on earth and by that means creates the awareness in which the current of earth intelligence operates. It is impenetrable to thought. No thought can pass out through it into the cosmic region of mind. Every thought or concept bounces back off the intellect enabling the individual to reflect, conceive and reason. Because of this reflection of awareness via the intellect all men perceive the same objective natural world, as do all the species.

Cosmic power holds constant the objects of nature and the earth we see from day to day. Cosmic energy originating from outer mind (outer space) enters the terrestrial mind picking up the ideas of the natural world. This idealised energy continues through the intellect across the psyche into the collective unconscious represented by the inside of the earth (or the physical body). There the energy is converted by the senses into sense-impressions of the idea and projected back to the intellect, which then reflects it in the individual's awareness as the physical world we all see.

The same intellect is used by the dead. But the way they use the senses is different. The dead live behind or inside the sense-mechanism; the living outside it.

THE POSITION OF THE SENSES

None of our senses is real. The whole of the senses, all together, forms one vital nucleus in the unconscious. That alone is real;

not the differentiated senses appearing as smelling/nose, seeing/eyes, hearing/ears, tasting/tongue, touch-feeling/skin etc.

Each sense has to be confirmed by another sense to be known. Take the example of sight. The eye cannot see the eye. The eye requires a reflective device (another eye) such as a mirror, or one of the other senses such as touch, to prove its existence. We say we see: this is undeniable. However, it is imagination to say that the eye sees what we see. It is impossible to prove the existence of the eye as an instrument of sight except by the negative means of obscuring, damaging or destroying it. But something cannot be proved to exist by the action of destroying it: that proves only that it does not exist.

Our concept of body-shape is derived entirely from the evidence of the senses. Sight and touch are 'positional' senses providing fixed perceptions. When sight and touch disappear, who can say what shape or form the body has?

So far I have mentioned only the five 'external' senses represented in the external organs of the body. But the sense-mechanism has more dimensions than the merely external. For example, the feeling of being hungry is a feeling that does not feel the body through the external senses. The body has feeling, or sensation, or being. Call it what you will, this sensation is its reality — its only positivity.

Our feeling is indeed inside the body. But our feeling body does not have the shape and appearance of the physical body that our senses or fixed perceptions would have us believe. Our senses have produced a body that ages and dies. Yet the sensation of the body never ages, never dies. And although sensation may alter in frequency between pain and wellbeing, it is never absent. This body of indeterminate sensation is the ground of our psychic existence, whether we are alive, dead or dreaming. This body cannot die.

It is always some other body which is seen to be dying. If I should ever see my body dead then I cannot be dead. If I never see my body dead then death is an assumption. Only our fixed perceptions are seen to be dying.

Here is another example of fixed perception. I am conditioned to imagine that this mind of mine is confined to somewhere in my head, when really it is not positional at all. 'My head' is one of its fixed perceptions. If the head is taken away, as in sleep, where is outside, where is inside? All is just space or psyche in which heads and dreams are created every moment.

How necessary are our senses, our physical bodies? The senses are utterly necessary to physical existence. But they are not vital to our psychic or non-positional being. Which means that without our senses or bodies, we still are.

Does this mean there is no physical world in reality? No. It means there is no reality in the physical world except that which I provide. If the physical world were real, no one would die and no object would disappear.

THE PERCEPTION OF REALITY

The external physical world is the projection of a real world onto the intellect, the world of ideas, which is then reflected in matter or sense. The sensory reflection is as real as any reflection an observer is prepared to identify with; and just as powerfully deceptive.

Inside the physical world, that is within the observer, is the vital, psychic world. And within the vital world is the reality behind all physical existence: the sun, moon and entire universe, not in the relatively flat perspective that we observe externally, but in its ideal, diamond-like splendour. From his place in the vital world man's double looks into this magnificent energetic world much as we look out at our beautiful natural world, the real world's reflection. And it is the double's vision of the real world that supplies the joyous feelings behind our perceptions of the earth's beauty. What we see 'out there' is devoid of such beauteous reality. Often we see the world as it is, in all its emptiness; and it is terrifying. Then we reconnect with the vital vision of the psychic double and although nothing has changed, life loses its drab, hopeless mundanity and is worth living again.

Thus the reality of the physical world lies first in the psychic world, in my double. In life, and especially in death, man can consciously perceive the supernal real world direct; but only as lucidly as he has managed while alive to free himself from identification with the physical world, and to that degree has united with his double. Until I can get control of my emotions I cannot discover this reality, or any other; I must go on dreaming and sensing it.

DEATH, THE REASONABLE REALITY

Both the psychological and vital roots of every man are in the world of the living dead, within him. It is self-evident that man's roots are not here on the earth. What is here leaves and disintegrates after each flowering — for it is man's lesser part though no less important.

His feelings are all within. His love is within, his thoughts are within, his pain is within; his satisfaction, ambitions and dreams are all within. The external world does not possess the reality each of us ascribes to it. We know this and live it every moment — through our immortal double. Otherwise we could not possibly go on living.

Of all the species only man has to live with the intolerable self-imposed concept of life with death at the end. A 'reality' that exists for perhaps eighty years is utterly intolerable; or it is tolerable because it is already unconsciously known to be a relative reality.

Death is just another theory, another partly observed fact that man has reasoned into existence using his external senses. To the rest of the species, death and life have no reasonable span and no reasonable distinction. Man himself only regards death as a possible end when it is happening around him, which normally is very seldom. Otherwise, he too lives confidently oblivious of it.

Man only resorts to reasoning when he does not know something. As soon as he knows he ceases to reason. Reasoning can solve many of his worldly problems, for in many ways it is a reasonable world. But man cannot use reasoning to know his own continuous, vital existence; it is a self-evident truth which he can realise, sense for himself and know at any moment. Like the rest of the species, he knows better about death — that it does not exist as an end. But because he does not consciously pause to realise that he knows better, he reasons perversely. His reason cuts him off from the truth, from awareness of his immortality, the ever-present feeling of his true being, his double.

Never has there been a moment when any man knew he did not exist. He may reason through recollection and memory that he did not exist at some time but this is transparent assumption and speculation.

His senses may tell him that someone he loves has died. His body may weep. Yet he still finds it tolerable and reasonable to go on living and dying in the same old way, generation after generation. This is not reasonable. It is illogical and insane. But the absurdity of it fails to penetrate reasoning man or unhinge the half-truth of his fixed perceptions about death or in any way remind him of the very thing that makes living tolerable — the fact that he and all things have permanent vital life.

THE FIFTH AND SIXTH DIMENSIONS

External space — the sense-perceived world — is defined as four-dimensional: everything in it is distinguished by length, breadth, thickness and time — or more simply, position. It contains no living or feeling thing; in spite of appearances, no object or person in external space has feeling or sensation. This may not be an easy proposition to accept.

109

Feeling and sensation are qualities exclusive to the fifth dimension which operates solely within the observer. Only one viewer/feeler exists in the world — the individual I who am viewing and feeling now. The contention that there is a plurality of viewers/feelers is an assumption based on four-dimensional reasoning; it has no truth outside of appearance.

Inevitably this raises the question of suffering observed in people and animals. The suffering exists only to the degree that it creates compassion in the observer — for I am the only one who feels pain or compassion. If the observer is unmoved there is no pain in the other creature, only the appearance of it — which is an assumption.

Compassion is vicarious suffering by one for another, the sole virtue in the sense-perceived world. Compassion alone has the power to end the appearance of pain in the world — the conditions in which our fellow man and the species are observed to suffer. In reducing the world conditions of hunger, injustice, cruelty, poverty and exploitation, compassion has the reflex effect of actually reducing the reality of pain; that is, pain that I the observer feel or will feel. This effect began to manifest as a fact of the physical world less than 150 years ago with the introduction of chloroform and the science of anaesthetics. The inspiration that led to the discovery of chloroform came from the sixth dimension, the world of ideas. Thus the world of ideas was able to reduce the pain of man in the fifth dimension.

This was the first real step towards alleviating man's physical pain and marked the degree of compassion that the human race had by then acquired. It exemplifies the higher self-regulating justice governing human existence. As only one feeling or fifth-dimensional being exists in the world — I, who am reading these words — and to the extent that I am compassionate and help to eliminate the appearance of pain in the world, I reduce the actual reality of pain for myself. This can be put another way. The reality of pain is the physical or emotional pain I actually feel. Although very real to me it has very little virtue or evolutionary value for me or for humanity unless it is endured vicariously (borne for another); or unless my forbearance and patience in bearing it arouses the appearance of compassion in another, so that they are moved to change the condition of the world that causes my pain. This quality lies behind the appearance of social progress.

The evolutionary outcome of compassion is to save myself from pain at some future date, whoever I may be. Therefore the future of man's pain, generally and for the individual can be safely predicted. As the conditions of pain (poverty, cruelty, exploitation etc) are eradicated from the world by compassionate action (a fifth-dimensional response) the reality of the pain that I feel will be progressively relieved by sixth dimensionally inspired ideas for scientific advances such as pain-killing drugs. Cosmic consciousness is also sixth-dimensional and alleviates suffering in the individual through higher knowledge.

Seven

THE
SEVEN LEVELS

*There are seven levels of terrestrial
mind. Each of them is embodied
in man and has its own place
in the structure of existence*

THE SEVEN LEVELS

THE EMBODIMENT OF MIND

AN IDEA CANNOT remain in the cosmic mind or human psyche without moving towards outer existence. The surge of the will is always towards manifestation. But the gap between the highest cosmic creative levels of mind and the human psyche in which the sense-perceived physical universe appears is inconceivably large. Any idea deriving from such an august, abstract reality as the sun has no means of manifesting in its original state in sense or matter. The character of the idea is too refined to allow it to become substantial, to make sense. Matter is too coarse a medium to respond in form and shape, rendering existence impossible.

By strength of will the power of the original idea is stepped down and held in four permanent states of deepening density. Original purity and simplicity are intentionally sacrificed to form ever more dense and differentiated strata of forces, energies and information. What began as unity degenerates into an actual world of simulated sense experience. Unity becomes duality, multiplicity and finally infinite complexity, so that the

115

idea which cannot really be, can be seen or sensed to be.

The power of the original earth-idea devolved into the terrestrial mind where it was held in four states. These are permanent states of mind, each of deepening density.

The four states or densities of terrestrial mind are:

First, the spiritual or cosmic state. This is consciousness supporting all intelligence.

Second, pure mental energy or intelligence.

Third, psyche.

Fourth, sense or matter, the state forming the material outer world.

Consciousness, the first state, supports and makes possible the other three. Inside consciousness is intelligence. Inside intelligence is the psyche. And the focal point of all three is the sense-perceived physical world.

CONSCIOUSNESS IN THE SEVEN LEVELS OF MIND

Within and supported by the four states are seven levels of mind. The state of matter supports Level One. The psyche supports Levels Two and Three. The state of intelligence supports Four and Five. The spiritual state, consciousness, supports Levels Six and Seven. In the space between Level Seven and Level One, mind becomes matter.

The sun's earth-idea formed the nucleus, the seventh level of the terrestrial mind. In the beginning it was all there was of the earth. The other six levels were added gradually, one by one, over what we must regard as an enormous stretch of time.

All seven levels of terrestrial mind are embodied in man.

Level One is his awareness of the outer world and physical universe.

Level Two is the subconscious.

Level Three is the unconscious.

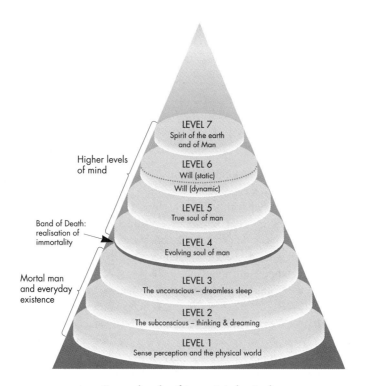

Higher levels
of mind

Band of Death:
realisation of
immortality

Mortal man
and everyday
existence

LEVEL 7
Spirit of the earth
and of Man

LEVEL 6
Will (static)
Will (dynamic)

LEVEL 5
True soul of man

LEVEL 4
Evolving soul of man

LEVEL 3
The unconscious – dreamless sleep

LEVEL 2
The subconscious – thinking & dreaming

LEVEL 1
Sense perception and the physical world

Seven levels of terrestrial mind

Separating Levels Three and Four is a narrow area called the
Band of Death.

In Level Four, as his psychic double, he is evolving towards
the ideal.

Level Five is his true character; ideal man.

In Level Six, as the will or great Being he upholds and creates
all the levels below.

At Level Seven man is the consciousness of the earth itself,
literally the spirit of humanity, the cosmic Lord of the earth.

The first distinguishing peculiarity of mind is individual

consciousness. That is to say, everyone in the physical world experiences themselves as I. This I persists in dreaming and at all levels up to the seventh and highest level of terrestrial mind where man is one with the sublime consciousness of the Earth Being.

LEVELS ONE, TWO AND THREE

Each of the seven levels is an actual world. This can easily be appreciated in the case of Levels One, Two and Three since everyone is familiar with them. The world of Level One, sense-perception, is the real world of existence. Level Two, the subconscious, has a different quality but we experience it as a real state of being we call ourself. In dreams we can feel the full range of emotions while performing impossible feats. At times the dream is so real we dismiss our own suggestion that we are dreaming.

Level Three is the place of dreamless sleep — another real state of existence. We enjoy it, look forward to it and usually feel better after it. We may not feel so good after a bad dream; but those dream images are in Level Two. Sleep without dreaming is as real a part of life as any, even though we may be oblivious in it.

We accept Levels One to Three as our normal experience. Normal is precisely what they are, as they mark the limit of normal or mortal man's experience. Normal man cannot pass beyond Level Three into the higher levels of mind until, while still alive, he passes through the Band of Death. Intrinsic to this transition is the astonishing realisation of immortality, that beyond any doubt he is the immortal I of individual consciousness.

THE BAND OF DEATH

The Death Band between Levels Three and Four separates immortal man from mortal man. To mortal man it is the ring-pass-not. All men are immortal but few realise it.

The passage through the Death Band is undergone while the individual is fully alive and active and brings psychological and/or emotional death. Physical death occurs well below it in Level One. In my realisation of immortality the death was simultaneously psychological and emotional but the two phases may occur separately. When they do, psychological death is likely to come first.

Psychological death is the realisation that death is an illusion; that it is irrelevant whether the body dies or continues to live. It is accompanied by intense feelings of love and beauty, but mainly it is an experience of pure comprehension and direct knowledge from within. Indirect knowledge received through the senses and the reason inevitably creates doubt and the need to know more, whereas direct knowledge from within dissolves doubt and eliminates the need to know. This raises enormously the individual's sensitivity to impersonal beauty, love and divine power. The psychological experience of death brings a vivid new permanent state of inner clarity which may be described as the realisation of space. Space becomes indescribably clear and pristine, enabling objects and facts to be seen in a finer and sharper all-inclusive perspective. This realisation represents the cessation of robotic naming and judging in the individual — the end of that chronic, conscious and subconscious habit of all men which perpetuates the existence of the false, superficial, intervening self.

Emotional death on the other hand is filled with sensation. It is the realisation of immortality accompanied by the full joy of release from the mortal burden of the emotional self. It is the realisation of the individual's oneness with all life and purpose whatever that might be. Both deaths are the outcome of a relentless search for truth or self knowledge. Emotional death is by far the more painful and personally distressing and is almost certain to include psychological death if that has not already occurred. It usually occurs in circumstances causing agonising emotional disturbance. This is of an intensity which can be equated with

119

the mystic's insane unrequited love of the divine, or with the loss or death of the most loved person in the individual's life. There is likely to be some awareness of having caused wretched suffering to others through the uncompromising stand of having put the search for truth before any other consideration.

I have been describing the extreme experience of my own simultaneous and sudden psychological and emotional death. For many people the passage through the Band of Death is a protracted and fragmented process of dying. All people at times throughout their lives touch on or enter it briefly. This occurs in moments of great personal loss or crisis when all is felt to be lost and there seems no point in going on. But the realisation of immortality I have described requires an extremely traumatic event and usually in a life that has been given to a relentless search for the truth.

LEVEL FOUR

There is no death in Level Four. Only in the denser Levels One, Two and Three is the drama of life and death enacted. Four is the level beyond the Band of Death where immortality is fully realised.

The realisation of immortality is probably the most splendid step man ever makes into the mind. Realisations of the other levels are undoubtedly more profound in abstract ways but the realisation of immortality is the first in which man gives the lie to his fiercest dread — death. Nothing can ever quite equal the astonishment of this first stupendous freedom from the universal lie.

At Level Four of the mind the living dead and the dying living contribute to the same evolutionary purpose, for Level Four contains the evolving soul of man. 'Soul' is a much misused word, now commonly used to describe some mysterious aspect of the personality. Historically 'soul' has been religiously attributed to individuals who demonstrate spiritual yearning, and by abstraction to some personal immortal essence that survives death. This conception is limited by ignorance of truth. There is no soul that survives death, only the impersonal consciousness which has no individual memory. No aspect of personality survives beyond Level Three. (This will become increasingly clear as you read on.)

The truth of the word 'soul' is that it expresses the principle and essential character of man. This principle operates in two phases: an evolutionary phase and a perfected or ideal one. The evolutionary or progressive phase takes place in Level Four of the mind. To distinguish the work of the soul at Level Four I call it the 'empiric soul'.

The empiric soul is what the individual man is now. It is not perfect. It represents in pure mental energy the result of endless earth experience of striving towards the ideal of the true or perfect soul. It also represents the evolutionary point that the individual man has reached now. Evolution is the attempt to reproduce in Level Four a replica of the true and perfect soul of man. If this were ever achieved man's life on earth would be perfect; the mythic Paradise or Kingdom of Heaven would have come to earth. Evolution would be at an end.

Just as the empiric soul evolves here, so also do the ideas which man uses and perceives. Take for example, the idea of table. No one knows precisely what this idea is. But its energy is constantly present in man's mind at Levels One and Two (where he thinks and perceives); and his attempts to reproduce this ideal table in all the tables he makes in the world are recorded as the single, evolving, formless image of 'table' in Level Four.

Level Four is the world of progress towards the ideal. Here occurs the first movement in the mind as each idea adjusts in shade and tone to the evolutionary changes occurring in the forms below. Here is preserved the real significance of changing fashion, style, preference and taste; all variations of the classical, moving in time towards the ideal — progress.

A man realising immortality is experiencing the state of his empiric soul at that moment. This does not mean he experiences himself as being any different from normal. How could he when his empiric soul is always what he is now? What is astounding and amazing is the realisation that what he is now is and always has been immortal. Yet he sees that he must go on living as a person whatever lies ahead. He may even feel like a lamb going to the slaughter, inasmuch as he must continue to live and die. But he knows it does not matter — wonder of wonders, he is forever!

LEVEL FIVE

The perfect soul of man is contained in Level Five of the mind. Here is the true soul of every human being who has ever lived. Each soul is a unique facet of the original earth-idea of man and all combine here — without ever having been separated — to represent one extraordinary, indivisible soul or character.

True soul is eternity. This true soul, man's eternal character, is the original earth-idea of Man. Being perfect, it never changes. Nor does it participate in existence or in Levels One to Four of the mind. Its role is static. It is a shining exemplar, the self-luminous image in which all forms of life and existence strive to perfect themselves in the trauma of evolution.

Level Five is the quintessence of man; the 'quinta essentia' or fifth essence intuited by ancient and medieval philosophers as a pure quality of being within the four elements of material creation, earth, air, fire and water. The ancient gnostics called it the divine body of the Anthropos, the one original Man suspended in eternity.

This soul of Man is realised at Level Five both as I the individual consciousness and at the same time as ultimate unity. In this realisation (and many have realised it) man knows he is one with all, that all things are in him and that he is in all things. It is an astonishing experience for I the individual to feel so personally privileged (and yet so personally insignificant or perhaps unworthy) as to be the very centre of this one indestructible, universal continuum of unlimited life.

The soul of Man sometimes manifests in the sense-perceptive world of Level One as the spirit of humanity, the living spirit. When this appears in a man it is the rare phenomenon of transfiguration. Without exception, to see it in man conveys the immediate conviction: 'This is the living God.'

Level Five may also be called the world of ideals. All ideas come out of this profound region of mind, even the idea of the soul of man itself. We could say that the ideas come both from Man and to man, since Level Five contains all the eternal ideas which man uses or draws upon in his evolutionary progress.

We saw how Level Four governs man's evolving attempts to reproduce 'the ideal table'. In Level Five the original idea of 'table' is permanently fixed. Level Five is the world that Plato evidently realised when he saw that every form has a timeless and motionless archetype in a reality beyond the world of the senses.

LEVEL SIX

Level Six is the world of will and impersonal love.

Will introduces power into the mind, creating the first duality. This duality divides Level Six into two halves: one static, the other dynamic.

The half closest to Level Five is the dynamic aspect of will — action. At this level, action is identical with all motion and movement in the world at any moment. When it is realised, this font of action is seen as universal or impersonal love. Love is the creating element in terrestrial mind. This inspires the insight that love is the Creator; that love is All.

Realisation at the dynamic level of Level Six is what is called 'self-realisation' in the eastern tradition. Self-realisation is achieved through a long cathartic preparation of right self-inquiry and meditation, right company, right instruction, right suffering and right self-centredness. Self-realisation is seeing through myself — the self I had previously assumed myself to be. It is the direct and immediate knowledge that everything spoken, thought of, and thought to have been done or achieved is the absolute ignorance of myself. Immediately the total ignorance of myself is perceived, I see through to the simplicity and nothingness that is the truth or God. Thus self-realisation is often referred to as God-realisation.

Self-realisation in Level Six invests all men who have it with the same unmistakable knowledge and authority even if they have never heard of self-realisation or Level Six, or been a reader of books. The truth is in the level of the mind, not in any description of it. Spiritual progress of any kind is possible only because each of the seven levels of mind exists as a real, separate, interacting world or division of knowledge. Without these levels there would be no way or stages into the truth-seeker's mind and all aspirations towards reality would be pointless.

The static half of Level Six, nearest to Level Seven, contains the will itself. The will formed Level Six out of itself by embracing the pure earth-idea of Level Seven. While at Seven there is idea only, in Six, idea is combined with will. Here in the static part of Level Six the will holds the earth-idea within a mighty potential of unexpressed, unstirring but infinitely knowing creative poise or purpose.

Realisation at this level is what the Hindu tradition calls the realisation of 'Purusha' — the single, ever-present, all-present, almighty Being, the unknowable original one. During my realisation of Purusha I was unable to stop laughing at the impossibility of sense or mind ever describing this mighty being that (in the realisation) I am. I suspect the eastern tradition of the laughing Buddha stems from this. This realisation is the realisation of the will itself and may be regarded as the highest realisation, even greater than self-realisation.

LEVEL SEVEN

Level Seven is the world of the first idea. It is experienced as pure, glorious spirit, perfection, the heavenly Lord of the blessed earth and the one body of humanity where no other exists.

Neither of the realisations in Level Six necessarily permits access to this pure idea of the earth at Level Seven. It requires the rarest of all recorded realisations and is possible only with the consent of the guardian will at Level Six. The will decides who is empowered, as surrogate for his fellow man in different ages, to realise the central idea of the earth and report back. This ensures that the truth and knowledge of the realisation never leaves the world of the living.

On realisation Level Seven is apperceived as Paradise, a literal garden of Eden identical with all man's most rapturous moments of love of nature combined in one single Being: as he has ever smelled, heard, tasted, seen and felt the earth to be in all its fullness; except that here the fullness is realised as a single unified Being free from differentiation into sense objects like trees, clouds, sea, sky and creatures, which in manifested nature inspire man's wonder and love. Moreover, this one originating beauteous Being of the earth does not vary or diminish. It is realised to be constant, divinely eternal, perfect in the unalterable substance of endless mind and spirit.

At this seventh level of mind there are no competing individual desires or interests to create war, conflict or pain. Only one Being is present, one earth, one impossibly sweet nature — the essence of the Being we call the planet earth, the one and only Man. This finally is what I am as earth man.

At Level Seven the earth-man consciousness ends. But although it is the end of a cosmic phase of mind, more profound levels exist as solar mind; perhaps requiring solar or cosmic man to realise them.

Eight

THE THREE PLANES
OF EXISTENCE

*Intersecting Levels One to Three of the
terrestrial mind are three planes of
energy. They determine the
qualities of existence*

THE THREE PLANES
OF EXISTENCE

IN THE BEGINNING there was nothing. Levels One, Two and Three had not been created. Levels Four to Seven were purely potentials and had no existence. Then, in the nothing appeared an extra-ordinary insubstantial plasma. This was the life force of the moon.

The moon appears to be a dead planet because its essence is being drained from it and used to produce life on earth. This is not an involuntary draining. It was and is a gift to the earth, a magnificent cosmic gesture of self-sacrifice by the moon whose selfless life-force is evolving as a result of it. In this aeon of existence, which eventually will be dissolved and refashioned by the cosmic will according to the law of evolution, the moon and the earth are permanently linked as twin planets. Through its votive offering as part of its own cosmic evolution, the moon one day will be a cosmic being supporting life and intelligence in its own right.

With the moon-plasma present, the potential higher levels of mind had 'something' to reflect on. Level Six, the level of will, focused on the plasma and impressed or characterised it with will and the idea of the earth. This transformed the plasma into the fundamental substance of terrestrial existence — the elemental principling factor behind the appearance of life and sense. Its name is 'vita'.

Vita is the first of three planes of existence and the stuff out of which the other two planes are formed. Vita has no external quantity or quality outside of the massed appearance we call existence. Just as the biologist's line between life and death can never be drawn outside of artificial parameters, vita cannot be identified. It can only be described and felt as 'life'. Vita is behind the flux of changing and vanishing forms. Whether decaying, eroding, disappearing, disintegrating or changing, life goes on. Vita contains all the force of unconscious matter; enabling it to explode, disintegrate and destroy whether through the exploding atom bomb or the devastating elemental fury of a hurricane or volcanic eruption. Fundamentally beyond human comprehension, vita is the power, purpose and mystery behind existence, matter and all the laws of science and physics. It is the indefinable essence of the sense-perceived world.

THE CREATION

The descent of the will onto the original moon-plasma which produced vita is recorded in the myth of Genesis: And the earth was without form and void; and darkness was upon the face of the deep (the pool of pure moon-plasma). And the Spirit of God (the primary attention carrying will and idea from Level Six) moved upon (impressed itself on) the face of the waters. And God said let there be light: and there was light. (And vita, the primary factor of light, form and all life was created.)

So the will descending from Level Six impressed itself on the moon-plasma and made vita. It then focused on the vita to form the first plane of existence. The colour of this plane is blue and so I call it the Blue Plane. This was the beginning of the creation at Level Three.

Next, from Level Five, another focus of attention descended on to the vita at Level Three. But Level Five, being inferior in power to Level Six, was only able to impress its characteristics of unity and the ideal on a smaller section of vita. This diminished area formed the second plane of existence; its colour is yellow.

Then the attention of Level Four, the evolving ideal, focused on the Yellow Plane. Level Four, again being relatively inferior, impressed a still smaller area which formed the third plane whose colour is red.

Although all three planes of existence are in the original vita of Level Three and continuously intermingle, each plane is associated with a level of the mind. Blue is associated with Level Three, yellow with Level Two and red with Level One.

Levels One to Three with their associated planes form the totality of existence. Everything that has ever been known or experienced, or ever will be known or experienced, must happen here. But the person we know ourselves to be exists only in the first two levels. As we dissolve the person through the more perceptive or spiritual life we begin to have access to the higher levels of mind through Level Three.

THE RED PLANE

The planes have particular significance because they represent the different qualities of existence.

The Red Plane, which is associated with Level One, generates feeling, which can only be felt in the sense-perceptive world. Fundamentally the world is perceived by man impersonally but a personal feeling or emotion arises as soon as he relates what is being seen to a past experience. The Red Plane therefore supplies the feeling of self, the composition of all the person's past feelings and emotions. As the energy of self, the Red Plane also supplies the energy that supports all attachments. It is the source of man's exclusive feelings of individuality. It is these feelings which give rise to all emotional attachments to people, objects, concepts and conditions. Thus billions of people are able to perceive the same physical world and yet by emotional identification create 'my' mother, 'my' body, 'my' house, 'my' country. It is this that produces our extraordinarily complicated and completely personalised world.

THE YELLOW PLANE

The Yellow Plane supplies the energy of intelligent reflection and corresponds with Level Two, the region of dreaming and thinking. Again, this energy is fundamentally impersonal and provides man's intellectual capacity. Even at Level One, if there is no personal identification with what is being seen and no reaction to it, then only intelligence is looking out of the eyes: the individual is looking into the world from an impersonal structure in the mind, the intellect. This is like having no film in the camera. The person emerges the moment there is any interpretation of the experience. This follows from mentally associating the event with previous (often subconscious) experience in Level Two by using the reflective Yellow Plane energy. The purely intelligent reflection becomes subtly personal as the individual creates associations of ideas through reference to personal memory and beliefs. So while the pure impersonality of the Yellow Plane can be expressed as 'intelligence without self', the yellow energy quickly becomes personalised, generates mental activity and in the complex of rational thought loses the pristine clarity of the intellect.

In most people for most of their lives, the Yellow Plane within their own being is persistently invaded by the energy of the Red Plane as selfish and emotional thought — fears, worries, doubts, anxieties, dreams, fantasies, speculations, opinions, attitudes and beliefs. This means that the part of their being which should be pure intelligence, a clear intellect, is constantly coloured and confused by personal emotional considerations. Personal evolution, however, enables the individual to gradually pull this invasive self back out of the Yellow Plane. When this is achieved the individual attains a natural equilibrium between the red and yellow energies. Then the self-perpetuating, habitual and continual judging, thinking and self-considering virtually ceases. The red energy is purified of emotionality and contained within the natural autonomic systems of the body. The instinctive impulse for survival — the source of the emotional self — is back where it belongs.

THE BLUE PLANE

The Blue Plane is unknowable and immeasurable — the unconscious. The energy of it is the end of the person, the destruction of everything human. When the human in the human being is destroyed, what is left is the utterly unknowable being. It is unknowable but is not beyond anyone simply being it. The Blue Plane is the basis of being and the ground of the senses at Level One. When the senses are purified of personal red and yellow plane energy (when the person is free of any emotion or self-reflection) the individual may have direct communication with the Blue Plane. This is so immediate, intimate and inextricable from life itself that to the individual it is his very being. The knowledge of this state has no mentation in it; nor is any awareness of sensation in the being dependent on the body's external senses — sight, sound, taste, hearing, touch-feeling. This 'knowledge' or 'sensation' is so inherently part of the being that it is best described as 'the sense of sense'.

The Blue Plane energy is both creator and destroyer, the power that decomposes and reconstitutes matter in the vast womb of life at Level Three. Nothing in existence lasts for long.

133

Everything returns to the Blue Plane and is dissolved there. At the very basis of creation is its own destruction. This power of the Blue Plane is mythologised in the Hindu triad of Brahma, Vishnu and Shiva who pervade existence as creator, loving preserver and terrible destroyer.

Nine

LIVING
AND DYING

*In this chapter we follow man's
journey through the planes of
existence in life and death*

LIVING AND DYING

THE AUTHENTIC BEING

WE THINK, DREAM and perceive in Level Two of the mind, the subconscious. But this is the tip of the iceberg. Beneath these images is our real being, what I call the 'authentic being in the world'.

The authentic being at every level is the intelligence we call I. It is our unchanging perceiving state, our permanently open window to the three worlds of waking, dreaming and the unconscious. Whether we are awake, dreaming, unconscious or dying it is always present and alert, impartially registering and recording all that is going on in all three worlds at once. Man (as we know ourselves now) is where his attention is at any moment. He cannot simultaneously be awake and asleep, or unconscious and dreaming. His attention is either on one world or in another, never consciously on or in two or all three together. But his authentic being is. Such simultaneity of perception seems impossible only because the timeless profundity and depth of being has not yet been realised.

The authentic being is the connecting thread along which our attention travels between the three worlds of the conscious, subconscious and unconscious — Levels One, Two and Three. It originates outside the three worlds in Level Four, passes down into the worlds through Level Three and ends behind the eyes in Level Two. Here it represents the mental aspect of our being which, combined with our feeling self or emotional aspect, becomes what we are at any moment — our evolving self.

Remember that the emotional aspect of our being is represented by the Red Plane. The mental energy of our Level Two aware-and-thinking self has a very close affinity with the Yellow. But unlike the pure energy of the Yellow Plane it now has experience in it.

The authentic being is the representative in the world of our Level Four evolving soul and retains the record of our evolutionary progress in this life, and in our life after death. When a man goes to sleep his attention travels back from Level One along the thread of his authentic being into the subconscious (Level Two) and/or the unconscious (Level Three). When he dies he makes the same journey, only the thread linking Levels One, Two and Three is snapped and his attention remains in Level Three. His authentic being is then the link between the red, yellow and blue planes of energy. After an important period of concentrated attention in the Red Plane he finally ascends up the thread through the Yellow Plane and the Band of Death to his empiric soul in Level Four — to the end of himself, and a new beginning.

THE SEX DRIVE

Contained in Level Three in the Red (emotional) and Blue (elemental) Planes is sex energy, the libido. Sex is the energy of death and the past — that is, instinct and emotion that has survived death — forever striving to perpetuate and express itself again in life. It is the energy of unconsciousness striving to attain to selfconsciousness. Sex, is the driving force behind all self-interested action. It is the irresistible pressure behind every thing man has ever desired or dreaded, forcing itself through his mental images at Level Two to colour and distort his view of the world and himself. It is the living past rising to impel us, often against our will and better judgment, to repeat experiences, good or bad, for the sake of sheer experience. Sandwiched between this awful pressure and the world is the authentic being, man's innocence. Its position is one of almost unbearable conflict.

From Level Six the will exerts an irresistible urge and longing for inner knowledge — for the authentic being to turn inward away from the outer world of experience towards the pristine higher levels of mind. But the impersonal elemental force and emotion from Level Three together exert an almost unbearable outward pressure on the authentic being struggling towards that deeper consciousness. The Red and Blue pressure drives men to madness while the counter-urge of the abstract will endeavours to prevent it. The hideous paradox of the will from Level Six fighting with its own creation, vita, symbolises the battle of the spirit against matter, soul against unbridled desire, principle against expediency.

Under the impulse of Level Five acting through the Yellow Plane the authentic being is under a constant though imperceptible pull to 'return home', to let go of the transitory physical world — to surrender identification with the emotional self's demands and rise to be united forever with the man's own divine true soul. But again the authentic being is powerless to obey its higher longings. Every man is tortured by doubts and fears about his life and self, feels insecure, inadequate and confused and longs for the peace he knows is possible but never seems able to attain. The element of self, or emotion, always intervenes.

Self is what keeps us from the realisation of our true soul in Level Five. The intensity of self is measured evolutionarily in our need to experience the world. While we yearn for the world we yearn for self and are Red Plane mortal beings. Inside each man's physical body at Level One is the mental 'body' at Level Two; and inside that, in Level Three, is the incursive emotional self that puts the pressure on man's authentic being, driving him into the world and almost out of his senses at times. The degree of self in each of us is where we are actually at in the evolutionary stakes. Until we are free of the need of the world, until we have overcome the world, not by force but by letting go of it, we must drive forever outward to experience what we lack and need. The force behind that drive is sex.

The moment of birth is the recurring restatement of our need to experience the world of the senses. From that moment on the relentless pressure of the emotional self forces the authentic being to identify with the physical world. It has no option but to register this misidentification on the embracing self. Closer and closer the emotional self presses against the authentic being until both seem to be one; and the ignorance is total.

By the time the rape is complete — coincident with puberty — the authentic being is convinced it is entirely dependent on the emotional self and the physical body for its happiness and survival. Death alone breaks the delusive bond.

FREEDOM FROM SELF

In deep dreamless sleep, when all our physical and emotional dependencies and desires disappear, we are in between the Yellow and Blue Planes and are perfectly content. Awake, we are seldom content for long. Even our dreams (expressions of self) frequently leave us disturbed and unhappy. Self, clambering for emotional expression through the senses and our dreams, is the main cause of man's discontent and dissatisfaction. It can never get enough of living and finally is baulked completely by old age and death.

While we are attached to material and sensory awareness — that is, while we live as though our physical body and emotions were all we consist of — we are unable to appreciate or tolerate a finer, less substantial reality. We cannot bear the peace and ease of Yellow Plane being. The lack of inner conflict and desiring becomes boredom. But when the need for physical and emotional excitement becomes the lesser part of a man, due to the realisation of life's impermanence and through having had enough of pleasure-and-pain living, he can gradually begin to withdraw into his Yellow Plane consciousness. Then consciousness, as distinct from selfconsciousness, is no longer felt to be the denial of life or living. When consciousness is free and emptied of self, I remain; and neither life nor death nor dreaming sleep can obscure my reality. The man eventually senses, and then knows, he is being informed in every moment of everything he needs to know about life and death without the necessity for emotional or physical participation to teach him the hard way. The hard way — alternating pain and pleasure — is the constant demand of a selfconscious existence.

THE PROCESS OF DYING

Death and living are the two halves of existence. Together they form one profound cyclic system in which the essential energies of all living things, after a distillation process, are returned to exist again on earth. This has given rise to the theory of reincarnation. But the error in the theory is that it implies personal continuity, whereas the purification process occurring in life after death eliminates every personal element from the individual's essential energies, so that what returns has no personal continuity. This is so complete that no one can say they have lived before. Nevertheless, there is a consciousness behind every living person that is completely impersonal — the authentic being. It consists of the value of all 'previous' lives. And that value determines the enlightenment or spiritual perception of the new person in the new life.

A person dying violently and instantly often experiences a temporary continuity of external awareness. Through the psychic sense of sense behind the now extinguished physical senses he sees what is going on at the fatal scene, even though he is dead. His awareness is sustained by a combination of two factors: first, the fact that his attention at the time of death was directed outward through the senses; and second, by the explosive release of his vital and emotional being into the space or psyche around him. This explosion away from the body into the surrounding psychic field is typical of all degrees of shock and, in extremes like violent death or the threat of it, provides the medium for a temporary psychic presence to be maintained outside the physical. If more than one person is killed in the same incident, or there are casualties or spectators, the intensity of the psychic field is compounded and the affected area extended. There can be two lots of people moving around the same scene — the dead and the living.

In the moments preceding a fatal accident the victim is usually extraordinarily unemotional, almost indifferent as he observes the succession of events leading up to the instant of his death. A kind of slow-motion detachment occurs. This is because the person's emotional self has already left the body and is charging the immediate psychic field. Emotionally he is outside himself. When self is dispersed like this the capability for feeling reaction is minimal and the authentic being, or I, registers the scene in images without feeling, that is without emotional interpretation.

In similar perilous moments of shock not involving death the authentic being can be externalised in the same way. Again, the dispersed condition of self reduces the degree of emotional identification and the person frequently witnesses the whole scene with detachment, this time from a point outside and above his own body.

Death of the body severs the thread connecting the authentic being to the senses; the externalised awareness diminishes and vanishes. The individual's distended emotional self gradually contracts to re-form his psychic body behind or inside the

142

senses. The person is then in the Blue Plane of Level Three, unconscious or in a dream state.

A person dying gradually and non-violently first withdraws into Level Two of the mind, the subconscious world of imaging and dreaming. This enables him to have glimpses of the forces of the Blue, Red and Yellow Planes of Level Three. His attention shifts between Levels Two and One. To an observer he may seem semi-conscious and rambling. To himself, he possesses all his perceptive powers although what he is perceiving may be beyond his mind's ability to comprehend or describe. He may simply lack the means, or the will, to communicate consciously to those by his side at Level One. Death occurs when the attention finally sinks into the Blue Plane, the elemental, universal unconscious.

The Blue Plane is a tremendously high energy source. One of its functions is to break down biological matter and extract all emotion. If experienced consciously, such primal energy provokes awesome, if not dreadful and terrifying perceptions. Even unconsciously its effects after death are disturbing. The living often experience it at night as a nightmare in which impossible tasks have to be performed or faced.

The Blue Plane is the reality behind our concepts of the physical interior of the earth such as Stygian darkness and molten core. From fixed perceptions of Blue Plane energy have derived the impressions of hell-fire and the sense of hopelessness attached to being confined in such a sunless underworld.

Blue Plane energy is the matter of the body, not its form. The function of the Blue Plane is 'essential'; it sustains the life-force intrinsic in all matter. The 'vital' (as distinct from essential) functions of organisms are secondary and are a Red Plane phenomenon. When vitality is cut off by the will through the Blue Plane the vital functions stop. Red Plane self has no power over life and death.

Each plane of Level Three contains a purgative or cathartic energy which provides the means for the dead person to evaluate as self-knowledge the experience he has accumulated in the life just ended. The Blue Plane energy disintegrates the body, removing every bit of emotional content from every cell. This appears externally as decomposition. However the inner process is completed in a much swifter time-frame. Respect for the dead and ceremonies and traditions that create an interval between death and destruction of the corpse are instinctive acknowledgments of the time normally required for the process to be completed. After extraction of all emotion, the remaining pure energy/matter of the body returns to the Blue Plane reservoir of earth elements out of which all forms, organic and otherwise, are ceaselessly being created. The extracted emotion combines with the emotional self which then passes on to enter the Red Plane for the individual to relive his life just ended. In this process he sees all that he did and didn't do, particularly for others, at a profound level never available to him while he was alive.

THE WORLD OF THE DEAD

The Red Plane is the world of life after death. Everyone who dies has a relative experience of the death world — relative because no dead person experiences it exactly as another does. Here on earth the emphasis is on a fixed formal world, the constant structure we all perceive. But in life after death the individual person creates his own world moment to moment; he literally enters his own emotional self. He does this by re-experiencing his past. But he does not see or realise that the world around him is himself, any more than before death he realised that the physical world is a projection from the Blue Plane within his own unconscious. His intelligence, instead of being at the centre of his own sense projections (the physical universe) is now at the centre of his own emotions. The position has reversed. Instead of looking out at the world and feeling his emotional responses within, his own emotions or feelings are what he actually perceives as the new world around him. At the moment of dying, the life he has lived will have unrolled before him without any emotion arising. This happens in the flash passage across the subconscious of Level Two. But now he is dead and in the Red Plane the emotional re-living begins. He feels he has only to think of someone or something to which he is attached for it to appear, or for him to be there. He does not realise, however, that such thoughts are prompted by his emotional condition and that he is completely controlled by this re-run through his emotions or past.

No one can ever hope to declare all that they are attached to. Often the strongest attachments only reveal themselves at the time they are threatened or severed. A person can seem completely detached from the world and unemotional, and yet after death know tremendous longing and need. Such longings and need are the fixation-power of the self; and it is the infinite possibility for the self to attach to things and conditions that makes the death world a uniquely relative personal experience.

Even though in the Red Plane after death each man creates a world of his own, he cannot disengage himself from the reality of the earth. Dead or alive, man's invincible, controlling passion is for life as it is lived, or may be lived, on earth. Living and death are two poles of the one complete earth-life. Living is the positive conscious pole. Death is the negative unconscious pole. The undercurrent in death, unconsciousness and sleep is always towards the earth, towards recurrence in the world of the living. The earth is the centre of all that man living or dead can know and wish for: there is no other existence he can aspire to.

While he is dead man returns to the earth frequently. But he does so in the negative mode of earth-life, via the unconscious. The unconscious streams down on to the surface of the earth from outer (inner) space. We, the living, perceive against or into the negative stream. Perceiving against the stream produces resistance. Resistance is positive and this is the positive mode of earth-life, our sense projection. Travelling with or in the stream creates no resistance; so it is negative. The dead travel down this non-conscious negative stream. With their negative perception they can see us; but we, looking against the incoming stream, are unable to see them. Stripped of our physical form — it is purely a positively induced phenomenon — we are essentially negative beings. Like the dead, the earth is a negative psychic structure. Only our positive perception makes it appear as it does. The dead perceive it in all its natural fullness, can participate in the elements by being the rain and the wind and are able to observe the affairs of the world or people to whom they are drawn — all without the knowledge of the living.

The dead person returns to that part of the world and to those things objectifying his strongest unconscious attachments. These may not be what he considered his deepest desires while alive. Once dead and in Level Three the man is freed from the Level Two subconscious and its superficial memory, which during his lifetime was responsible for most of his explicit attachments and concerns. These may now have no meaning at all for him. In death man has entered the deepest unconscious part of his memory where his true attachments and desires reveal themselves.

His first experience in the death-world is of the place, object or person closest to his heart. The people he experiences may be alive or dead. If they are alive he will perceive them as they are on the earth. If they are dead he will be united with them in the world of the dead. If another dead person was an enemy, or the victim of an action which caused intense guilt, a confrontation will occur; but without any overt conflict outside of what the dead man himself registers. This can be a violent experience for him, depending on the force of emotion involved; but inevitably it is salutary.

Anything a man wants or is drawn to he can do, as long as it does not require a conscious earthly response or the sharing of other dead people's emotions. Freed of the body he can indulge every desire he ever had, except for sex; for that he needs a physical body. If he wants to fly over oceans, cities or other parts of the earth, he can. In this he can be described as the past flying over the present. He can perceive the earth exactly as it is and yet have no effect on it except through living people and their emotions. He can move freely among the living but to make his presence felt he must use emotion or love.

Only through emotion and love is it possible for the dead to communicate with the living. Emotional communication, being a projection of self or desire, has a wide range of possible effects on the living, most of which produce some kind of negativity or apprehension. The dead in their ignorance, or by their desires, can produce moods of gloom, fear and confusion in the living, according to the emotional connection they are able to establish. Most living people lend themselves to the errant dead by being emotionally cooperative through excitement, brooding and over-indulgence in sex and sexual fantasising. But love or affection for the living has a different effect. These positive communications are mostly a mixture of love and emotion. They allow the recently deceased person to project a sign of presence to a living loved one, conveying a definite feeling that contact has been made, without arousing anxiety.

The returning dead perceive people and events on the earth as they are in the present. But to them the present is not the crucial point of existence that it is to us. For them the time element has changed. It has lost its narrow focus, its urgency or edge. This is because there is no future in the death-world. The future is only a device of the living for trying to put off facing the fact of death. When death removes the future all the strain goes out of the present or the person.

The death-world consists totally of past — the surviving vitality of all life that has ever lived on earth. The past of the death-world is the living substance of the future. That is, there is no substance in our future apart from death. This is the most important lesson any consideration of death can teach us while we are alive. If we can see the truth of it, even glimpse it, the strain of the future immediately starts to go out of living.

As the future among the living is only a configuration of the past, the dead if they wish can perceive our future for us. They have only to look into our past, and there it is; but the future as we understand it is of little or no importance to them, outside of any concern they may feel for the problems of their loved ones on earth, with whom they are still emotionally linked. Whereas the living are pre-occupied with trying to secure or fill in their future, the dead at this stage are preoccupied with reliving their past.

PURIFICATION

Having relived his past and exhausted his interest in it, the person now enters a new state of consciousness. Before, he had a knowledge of the people he'd loved and the objects on earth he was attached to; this is now all gone. Before, he was still personal; now he is impersonal. He has literally withdrawn into the centre of himself to become his pure ever-vigilant and conscious authentic being. This point of consciousness contains the value of all the lives he has ever lived, but it has not yet received the value of the life just ended. This process now begins.

His consciousness passes from the centre back out through the emotional self towards an outer ring of abstract energy. This is the Yellow Plane. On the passage through his emotional self, instead of enjoying the freedom and the pleasure as he did before while passing through the Red Plane, he now evaluates the worth of the life just ended. That value is then added to his authentic being and he unites with the Yellow Plane energy. His old emotional self dissolves and disintegrates, the discarded pieces falling back into the reservoir of constantly recycling Red Plane energy.

The man, as his authentic being, has now gravitated to a setting where some form of creative work is begun. This work is his endeavour to assimilate the value of his past life and transmute it into something finer — what he could have been instead of what he had been. By drawing upon his virtues to balance his shortcomings he is able to adjust and amend the life just ended — in short, to rectify his past. Death here provides this unique opportunity.

Having already reduced his whole existence to the abstraction of the Yellow Plane, the value of his life becomes the work of his life. Here after death every being generates some form of artistic expression which he works on with great seriousness and immense satisfaction. The work avenues available on earth are extremely limited by comparison. Whole new ranges of self-expression open up for him such as may be imagined to exist between a garden-lover and a sculptor; or a school teacher and a philosopher; or between a man who just loves to walk and an athlete or a dancer.

Gradually the being becomes completely absorbed in his work. For a time he can be called back to the earth by clairvoyants or by living loved ones with their emotions. But as time goes by he is too absorbed in his art to be reached from the earth or be bothered with it. What the dead man probably will not realise is that the work he is enjoying so much and slowly disappearing into is the designing of his next incarnation. With the materials provided by his last existence he now labours to produce the finest work of art he can manage — the outline of his next life on earth. This outline will be modified by a number of other important factors, namely: the discrepancy between his evolving soul and the true soul; external cosmic, solar and planetary influences; the influence of the Blue Plane vita out of which his new physical body and environment will be formed; and the influence of the Red Plane instinctive energy that will go to make up his new emotional body.

When the dead man has used up all the materials at his disposal he rests content. He can do no more. In this state he can have no fresh desires. Being his own authentic being, the ageless empiric I of consciousness, he encompasses the abstract sum of all his labours. He is almost pure being, pure abstract self-knowledge. He has virtually dissolved himself and is now 'being' as nothing, reduced to minimal existence in the Yellow Plane. Since there is no longer any self to be conscious of, he is no longer selfconscious. He is now in fact conscious. But unless while alive he had begun to identify consciously with this self-less state, his knowledge of it will be identical to deep dreamless sleep. If the man could only be conscious of his state, content in his being and aware of his I without self, he would not be asleep. He would be real or objective — self-sustaining, complete; and would have no need ever to 'wake' because he would know that whatever he awoke to would be some sort of dream, a substitute world or time. But as he did not learn to give up the need of the world to mirror or substantiate his existence while he was alive now it is as though he were in deep dreamless sleep. The man is really awake but does not know it. So he lingers in a kind of expectant condition waiting to wake up when he is already truly awake.

What happens next is the I or authentic being passes up through the Death Band into Level Four. The Death Band dissolves the last vestige of vita from the authentic being. It is cleansed of all trace of existence. Even the infinitely faint self-consistency of the Yellow Plane is removed. The authentic being is now pure mental energy. On reaching Level Four it unites completely with the individual's evolving soul. The union is registered as a change in the soul's vibrational frequency or colour: the immortal being of the former person has then evolved that much more.

The evolving soul at Level Four is itself an incipient art form, a kind of energised being forever seeking the means and inspiration to fashion itself in the image of its creator, its true soul in Level Five. This endeavour is true art and is precisely what art is about in the external world. But the art a man expresses in any one incarnation is often minimal compared with the innate artistic capability he has evolved to. His soul is itself the art-work of many lifetimes that may or may not require him to appear on the earth next time as an artist. He can have the capacity of a great artist and yet express this in some way quite unrecognised by the world. Man is the artist, man is the materials and man is the art. One earth life is but a few brushstrokes sometimes revealing a high point of colour or drama in the never-completed picture of a man's soul, but very often signifying simply the filling in of some essential though unspectacular piece of background.

When a man's authentic being merges with his evolving soul, that particular life and life-after-death are ended. No trace remains of that particular individual in the worlds below. That person who lived before is never seen again. But immediately the man — not the man he was but the man he now is — comes again in a new form. From his evolving soul in Level Four a beam of attention descends on to the Red Plane of Level Three. Passing through the Yellow Plane it picks up the outline of the man's next life which he had left there and projects it onto the Red Plane reservoir of self. Implicit in this radiant beam of potential life is the man's authentic being. Immediately, like iron filings attracted to a powerful magnet, Red Plane energy clusters around the focus of the new life to form a new emotional body. Although consisting of past or self, this energy is not necessarily the individual man's previous past or self. That is all held abstractly in the man's authentic being. This Red Plane energy is out of the whole reservoir of past or self left behind by all men and life during the after-death process since time began on earth. It is intensely instinctive, passionately attached to vital existence. In life it provides the instinctive knowledge which ensures survival of the organism until the last moment; it is the living vital memory of the species and the race which is intrinsic to the unconscious of all men.

The new man now consists of a highly energised centre still waiting to awake, and a surrounding vital or instinctive body. He now exists as almost pure need and emotion. Rising as a wave of emotion from the Red Plane through the Blue Plane, he spirals up across the subconscious of Level Two (through the dreams or thoughts of the mother) towards rebirth and personification of the experience and self-knowledge he lacks.

His emotional body is shaped something like a wire spring coiled into a cone with a single point at the top. This is the point of entry into the psychic womb of the mother, which to external sense-perception appears as the physical womb. His entry is equivalent to conception. From here the new person is finally born into the world at Level One.

LOVE IN THE WORLD

Before man, the earth was devoid of love. Beauty as nature alone existed. But nothing in nature could reflect on that beauty. The power to reflect on beauty came with man. And out of that emerged the power to love. Man has been struggling ever since to bring love and more love into the world.

It is every man's unrealised longing to match the beauty he senses with the love he feels. In the span of his lifetime this is as difficult to do as to understand because each man has to create that love out of his own emotion. Emotion is not love; but it is all man has to start with. The whole point of living, dying and the re-incarnating process is to imbue man with an ever-deepening sense of love. Through all his lives of pain and pleasure man grows in understanding and the ability to convert his personal emotional self into impersonal, non-emotional love. The dark red energies of self are purified and refined into ever lighter shades or frequencies approaching the finest of pinks — love in the world.

As already described, after each incarnation is completed and before that particular person has finally been dissolved, a man works assiduously at designing his next life. The skill and artistry (or otherwise) he employs in this handiwork depends on his power to love. If emotion has been his highest form of love his efforts (as his next life) will reflect this. Emotion is only love unrealised but it reproduces self-love and self-service and therefore a limited, unfulfilling life.

However, even love is not an end in itself. If it were, love would be the cessation of action. But love *is* action. In its finest shade of pink it becomes action free at last from self-consideration because the direction it takes is invariable, one-pointed towards bringing something even finer into the world — character.

MAN'S CHARACTER

As the Red Plane is the source of man's emotional nature, so the Yellow Plane above it represents the seat of character — his spirit or true soul. Man's emotional nature is a condition of selfconsciousness, therefore a limitation of his being. He feels bound, restricted and sometimes threatened by it. Character on the other hand, is a state or quality beyond selfconscious knowing. Man does not feel restricted or threatened by character. While he invariably admires it in another, he can never be aware of it in himself.

What we see in ourselves is not character but our own nature. If we think we see character in ourselves we are selfconscious; what we are seeing is our emotional self or a misplaced notion of our self. Being unknowable, character can only appear in us when we are not selfconscious. Although it appears in individuals it is not individual; and it never changes. But seated as it is above the denser Red Plane it gets very little opportunity to shine through man's normally fluctuating emotions and the superficial demands of his selfconscious personality. For character to come through with its unmistakable power of presence the polarity of a man's nature has to be altered. It must be given a negative value in place of its worldly-wise positive value of self-centred projection and consideration. Developing love does much to achieve this transformation but ultimately self has to be dissolved sufficiently so that consciousness is the greater part of his being.

Sooner or later every good man who has learned to love and perhaps to serve something worthy in the world reaches a crisis point. He has to start facing up to the only thing that now stands between him and true selfless love. This is himself, the final residual red-raw emotional entity which he overlooked and unconsciously assumed was loving or doing the loving. The self itself — representing all that lives in him apart from love — must now be consciously dissolved. This is done by using his authentic being to observe and understand his emotional self. It means turning his attention inward instead of allowing it to be driven outward. For a long, long time the man must live out of character instead of out of his nature. This is extremely difficult to do as it entails going against much that is psychologically and emotionally normal in himself and society. To all other men living normal lives his behaviour and actions at this time will seem to be unnatural and therefore questionable. Furthermore, while the struggle is in progress the man's ability to love may seem to the world to be doubtful or self-centred.

By persevering a man gradually aligns his nature with his character. When the alignment is distinct enough the character shines through, giving a new dimension to the natural being. A man can then be said to have had power or presence added to the force of his nature — and this is equilibrium or willpower.

Character is not commonly seen in the world. Even when it is seen it has no discernible continuity like the nature of a person or his personality. If personality is to be termed positive, character has to be described as negative. Character is not so much seen for what it is as for what it is not. That is to say, character usually appears or stands out in situations where the normal instinct of self-preservation or common sense is to go in the opposite direction. Character tends always to be isolated, to have to stand alone or out front. Also it tends to prove of very little value from a worldly point of view apart from the virtue of its own existence for anyone perceiving it in another.

To observe character in another usually evokes compassion, spiritual love, or the sense of gratitude. Gratitude is the finest of all emotions, but it often deteriorates into a form of self-service with the giving of gifts in return. We give gifts to assuage our selfconscious feeling of not being able to give enough of our self. To be worthy or evolutionarily effective, gratitude demands first an immediate offering up of the feeling itself to the 'unknown' within, the one and only character at Level Five above. It is then divine gratitude. By expressing gratitude first to the source of character and not to another human being, selfconsciousness is helped to be cleansed of self. Any giving of gifts can come later as a complement to the offering up of gratitude.

Character and nature are a world apart. Character is always present and implicit in being, as beauty is implicit in life. It does not incarnate. What incarnates are the vital, self-projecting evolutionary needs of man's nature. This may seem a deplorable constraint for the world. But the justice of it is undeniable: it is up to each individual to express character and virtue in the world by manifesting them himself if he thinks those qualities should be here. Otherwise he — and the world — is stuck by default with his narrow emotional Red Plane self and existence.

Ten

HIGHER
INTELLIGENCE

*How can scientific man make
contact with cosmic
consciousness?*

HIGHER INTELLIGENCE

THERE IS A rough correspondence between the forces of matter that science works with and the psychic origin of the forces. In science a molecule is defined as the smallest portion of a substance that still retains its chemical identity. In the psyche, the Red Plane consists of spiral-shaped 'molecules' that make up man's chemical identity, his reactionary self. These are molecules of self or instinctive past, the emotional stuff people are made of. They are mutant and influenceable. As molecules are larger than atoms and immeasurably larger than elementary particles, so Red Plane molecules of self are larger than the atoms of the Yellow Plane, or man's spiritual identity, and immeasurably huge compared to the elementary particles and forces of the Blue Plane.

So far science works only with the energies of one of the Planes — Blue Plane vita, the substance out of which the other two planes are created and which contains the elementary forces and particles of physical creation. Here is the source of the physicists' sub-atomic particles, electro-magnetic radiation, anti-particles and the endless variety of tangible and abstract forces which meet and play at the edge of mind and matter. Study of Blue Plane energy provides an understanding of matter but in pursuing it alone and ignoring the inner world of the scientist himself, science remains a materialistic discipline.

THE PERMANENT ATOM

Atoms of the Blue Plane with which the physicists work are not permanent. That is, they can be broken down by the mind into more elementary particles or energies. There is no end to these particles, it seems. Even among the physicists there is a joke that whenever they are stumped by a fundamental gap in their knowledge the gods decide, 'Let's throw 'em another particle'.

Eventually a new generation of scientists will discover the permanent atom. Permanent means causal. The permanent atom is the constituent energy of the Yellow Plane. It is a unit of supreme intelligence, freestanding in the mind, a combination of will and idea. Its quality is I. It has no material existence but it does have integral existence, which is existence independent of matter, whether in or outside matter. It can also be called 'presence'.

When the scientific consciousness reaches the level of abstraction needed to register this presence, a scientist will experience or discover the permanent atom. He will hear a voice of unquestionable authority addressing him in his mind; a voice from either inner or outer space which will shake his whole structure of scientific and personal certainties. But unless the individual man who is the scientist has been adequately prepared by catharsis of his Red Plane emotional self, direct communication from Yellow Plane intelligence may cause an emotional breakdown. I suspect this may have already happened. The first scientifically acceptable communication may take place simultaneously in the minds of several leading physicists or astrophysicists. If they have the courage to reveal and share the knowledge it will at least provide them with some reassurance about their mental states, and hopefully encourage a determined and concerted effort to identify the source of such intelligence.

The permanent atom or I is behind all intelligence including our own. For us it can be said to have three aspects. These three aspects can be called the higher-mind point, mid-mind point and lesser-mind point.

From the higher-mind point, the permanent atom or I contains all there is to know without any need to reflect on what it knows. It is completely self-consistent. This is the realm of what would be called infallibility or omniscience. From the mid-mind point, the permanent atom represents the intellect which has a reflective role; it cannot help but reflect the world in front of it. This reflection creates in the mind the phenomenon of existence. From the lesser-mind point, the permanent atom represents the memory function, again a purely reflective role. As the memory it reflects stored sense-perceptions (individual impressions), thus creating an intellectual sense of continuity as self. This is the realm of ordinary living and relationships.

The negative quality of the permanent atom places it outside the speed of light. It is omnipresent. Neither its speed nor position can be measured; it has neither. This is because its motion or velocity is integral. Integral velocity is this-moment which has no time (for reflection). Terms like instantaneous and simultaneous are inadequate to describe it as they come under the inferior timescale of sense-perception and light.

Presence of mind

From the higher-mind point the Yellow Plane and its permanent atoms is behind all events. The source of all action, it never moves, remaining poised beyond and above all the confusion and self-serving which characterises human existence. As the highest point and end of Level Three — of existence — the Yellow Plane is suspended over the Red Plane of self and confusion like the sky above the earth. However high man ascends, it is always there. In praying to his god or gods, lifting his eyes to heaven for a sign or answer, whether sighting UFOs or just contemplating the night brilliance of the universe, he is acting out and reaffirming his timeless recognition of the Yellow Plane heavens as the source of his faith, hope and inspiration.

'Up' in the Yellow Plane, away from self and confusion, the mind of humanity divides into two divine presences of higher intelligence. These are the devotional presence of mind (consisting of 'left-spin' permanent atoms) and the scientific presence of mind (consisting of 'right-spin' permanent atoms). The devotional presence of mind is responsible for all artistic and loving urges: giving or the attempt to give. And the scientific presence of mind accounts for all intellectual activity: receiving or the attempt to receive as the pursuit of information and knowledge of any kind. Anyone can observe these two basic motivations in his psyche for they determine which one of only two possible directions every thought, aspiration and activity will take. Every moment the individual is a reflection of one or the other, or an admixture of both. Since all ideas coming down to man through the Yellow Plane are given left or right-hand impetus, these two presences are behind every conceivable effort on earth.

The two divine presences of mind contain the nature spirits of myth; the genii and muses of music, poetry and other arts; and the gods of war, providence, and other deities associated with various skills and disciplines. The development in the ancient world of the skills of war together with the processes of government, politics and administration arising from them represent the first attempts to reach (or return to) the scientific presence. Religion, mysticism and art were endeavours to reach the devotional presence.

At the summit of the mind of humanity is the transcendental presence. This is the presence of the earth spirit itself at Level

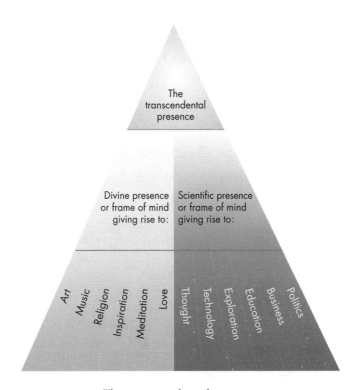

The transcendental presence

Seven descending through Level Five to inform the atoms of intelligence in the Yellow Plane. All intelligence on earth derives from this single supreme presence. Out of it devolve the dual devotional and scientific presences below and out of those in turn, the lesser spirits, deities, beings and selves, the whole determining the hierarchic structure of the human mind.

The transcendental spirit together with the two lesser divine presences forms a divine Yellow Plane trinity. Each presence in the trinity is a spiritual state which the individual can realise in himself as an all-knowing, all-present intelligence such as would be described as a divinity or God. But realisation is not the point. The summit of man's evolution is to be that indescribable being.

The Yellow Plane stands for life beyond death — whereas life after death is represented by the Red Plane, as we saw previously. Life beyond death is another way of talking about the cosmic participation which occurs at the extremity of the human psyche, in and through the Yellow Plane at the beginning of the abstract levels of mind. Any visiting extra-terrestrial intelligence has to come down through the higher levels of mind and finally express itself through the permanent atoms of the Yellow Plane.

COMMUNICATION FROM HIGHER INTELLIGENCES

The permanent atoms of the Yellow Plane are constantly informing mankind from above. Their presence is recognisable by the unique quality of their information: originality as expressed through truth, science or art. But as the knowledge has to come through the selfconscious and emotional Red Plane, much of it, even in the highest expression, is lost or unperceived. Or it gets twisted and distorted.

While the Yellow Plane is responsible for every new and original idea man receives, the Red Plane repeats the idea and generalises it. The important point is that Yellow Plane communications are received via the self and to the degree that the self has not been cleansed of past, the clarity, newness or truth of the message is reduced. Also, the self thinks it got the idea or invented something when really it was only the receiving instrument.

When self is absent there is no doubt that the message comes from a higher source. Direct communication from the Yellow Plane is the experience of personal contact with an independent and far higher intelligence than the limited self of any man. Selfconsciousness has then been superseded by consciousness.

Typically, the 'voice' of God, or a god, or of an intelligence from another place, planet or star system is registered in the mind distinctly, and the recipient will have no possible doubt in that moment as to its truth and validity. But all sorts of obvious difficulties close in which make such communications extremely dubious, not only for the man concerned but also for the world which he may then try to convince of the validity of his experience — only to end up desperately defending the soundness of his reason. Immediately after the experience, due to an inrush of self or emotion, the man may lose his clarity and certainty and start thinking he imagined it, or doubting his own sanity. The world itself has good reason to doubt this kind of experience which is common among the insane, religious manics, drug-takers and lower psychics all of whom, due to Red Plane or psychic distortion, are confused to begin with and unable to communicate its simple truths.

However, genuine direct contact with higher intelligence has already been successfully and permanently established in one area of human consciousness. This is maintained by what are called God-realised men, the genuine spiritual teachers who have realised at least the sixth level of mind, the originating Higher Being or Purusha. At this level and above 'voices' are no longer necessary; the knowledge comes direct out of the teacher's Yellow Plane consciousness.

Below the level of God-realisation permanent direct contact with higher intelligences is not possible to sustain. After a time they will seem to break off contact and just vanish. The danger in this for an individual hearing genuine cosmic 'voices' in his consciousness is that inferior psychic forces will rush in after them and fill the vacuum with misleading information and false revelations. Joan of Arc and many others in history and our own time have found to their despair and consternation that the voices they had learned to trust suddenly became unreliable.

The reason for all this is that higher or cosmic intelligences have no past whereas our earthling sense-perceptive awareness depends completely on past. We see and hear through the accumulation of this past which is the space or psyche around us. Our internal apperception and conceptualising ability also depends entirely on past — only in this case the inner accumulation of that past is our self. Self and past are synonymous except that self is the past or space within, in which the individual thinks, while past includes the outer sense-perceived space in which the body moves.

Since cosmic intelligence possesses no past it has no space/time of its own in which to manoeuvre or communicate. In order either to appear visibly as an object or to communicate as a presence in the mind, it must use either the earth's external past or space (the psyche) or the individual's interior past or space (self). In doing this it actually 'uses up' the past or self supporting it. A phenomenal visitation such as a UFO requires and consumes far more past than a one-to-one psychological contact such as a voice or presence in the consciousness.

When the particular stratum of past or self being used is exhausted the contact cannot continue. On the other hand, any man whose self or past is sufficiently dissolved transfers automatically to the level of cosmic consciousness. Here he is permanently in phase with every other intelligence on his particular mindwave in the immediate universe. Each individual has a unique oscillatory cosmic frequency with which he can 'perceive' — once attachment to the psyche or the world's pseudo-reality is broken. This book, for instance, has been written on a cosmic mindwave.

UFOs and the speed of time

The rate of consumption of past by manifesting cosmic intelligences is extremely high. Consequently, UFOs appear to move at great speed, and to communicate in the mind almost at the level of illumination.

The UFOs which millions of people have sighted since 1948 carry out a cosmic task to assist the planet's evolution by burning up vast areas of past that the earth and the human race have no further use for. Without this destruction of past by these cosmic forces of the present, time for us would have slowed down. Too much past would have accumulated and choked the psyche. However, far from slowing down, time in the form of man's comprehension and communications started to speed up. But as the speed of communication increased there was more for man to think about. This resulted in a burgeoning of past which will continue as man's electronic wizardry proliferates. Sightings of UFOs will multiply and other cosmic manifestations may possibly appear.

The chief effect of swifter communication for the scientist was to make the speed of light an inadequate frame for his theoretical calculations, even though it continued to remain an absolute in his sense-perceived experience. The changes this speeding up brought about in scientific awareness as a whole were so bewildering that specialisation became the only way for scientists to preserve a grip on the almost contradictory diversity of knowledge. Specialisation will increasingly become the scientific way until ultimately specialists linking the specialist fields will be essential. Meanwhile, the masses will become further and further isolated from the esoterics of scientific knowledge while at the same time making more and more use of its revolutionary practical products. Finally, it will be realised that through the products the scientific mind is using the masses instead of the masses simply using the products. This will be a period of extreme anxiety among the more intelligent people on earth.

For mankind as a whole the speeding up of time, especially in the second half of the twentieth century, has been responsible for the rebellion by younger generations against the old values and hypocrisy of assumed authority. Attempts to manifest more love and understanding in the face of authoritative violence and indefensible war, although many of them short-lived, combined with other liberating influences to produce a vast movement towards tolerance, self-discovery and alternative education.

THE HUMAN CHARADE

Since the earth and therefore the individual man is ultimately a cosmic being, every member of the human race has the potential in time to participate in the one cosmic consciousness shared by numerous higher intelligences in the immediate universe. However, at the present level of our development we are still well and truly earth-bound. So much so that in spite of our interplanetary and other space explorations we are in effect going nowhere. We are exploring an empty universe. We will not find or encounter any other intelligent life out there except perhaps in the same exasperating and inconclusive way that UFO sightings suggest the possibility of higher intelligences observing the earth.

The entire compass of man's experience, all he can ever discover about the planets, sun and other stars and star systems, is contained in the Seven Levels of terrestrial mind. Each of the levels of pure mind, beginning with the Yellow Plane in Level Three and continuing in the higher levels, is a rising degree of cosmic consciousness. At Level Seven the consciousness culminates in awareness of the cosmic or universal Earth Being, the earth-idea itself. For humanity, still enmeshed in the psyche way down in Level Three, this has a very odd and unsuspected effect. It means that all the scientist discovers about the planets, the sun and other stars is already contained in his own unconscious, the terrestrial mind. He (and for that matter all of us when we discover anything) merely brings or subconsciously recalls this knowledge up to the surface into his own psychic awareness, surprising and delighting himself with what he already unconsciously knows. This makes our whole existence completely subjective, tantamount to a charade in which we unknowingly and unconsciously act out each day the discovery of knowledge we already have.

The human psyche in its present distorting condition cannot know the planets as they really are no matter at what close quarters the investigation is made. While man continues to search for significance only in a sense-projected, external universe he will discover a second-hand reflection of reality, not reality itself. The planets that the scientists are investigating are only symbols, and the significance of any symbol is always in the unconscious of the viewer. Whatever man deduces from his studies of the planets will necessarily be a past concept — the symbols or planets merely reflecting what he already knows and what he knew unconsciously he was looking for and going to find. The result is always materialistic; that is, it confirms his laws and theories giving rise to more laws and theories but never revealing what is really there and new — higher intelligence.

Intelligent life abounds in the universe around us irrespective of science's conclusions. What the scientists are looking for on other planets is terrestrial life; cosmic life is there, but they cannot see it. Science has even established that its terrestrial laws of physics apparently operate at the edge or beginning of the universe — a fine example of the charade — but scientists still have not seen one sign of intelligent life in all that infinity!

As man is not cosmically intelligent himself he cannot realise and cannot understand that the only purpose of space exploration is to make contact with higher intelligence. But he cannot make this contact because all he gets back is the level of intelligence he is himself employing — intellectual materialism.

Cosmically, man is an earthling. In the limitless cosmic interaction of universal mind or intelligence, earth status merely means he has a cosmic obligation. At present this obligation is to get his humanity on earth right. He must evolve a global system, a civilisation, which eliminates the current necessity for injustice, hardship and poverty — the fruits of intellectual materialism. He must do this by taking responsibility for all life on the planet and not just his own self-interested position. In the meantime, no matter how many individual men realise cosmic consciousness there can be no participation by humanity as a whole in cosmic or extraterrestrial affairs. Cosmic consciousness reveals or realises this, which is why every man reaching cosmic awareness will be found involving himself with the immediate, obvious plight of humanity as he perceives it.

Humanity cannot join the cosmic club of higher knowledge and intelligence until it fulfils the basic condition of membership: it must get its own house in order. The blight of a consciousness which can tolerate the conditions under which most of humanity has to exist cannot — will not — be permitted to spread beyond this planet. Intellectual materialism and the Pandora's Box of miseries it has released in its long evolutionary trail must be confined to earth. The learned doctor must learn to clean up his own mess; no orderlies will be provided.

No matter what man accomplishes in space exploration, the human or public interest in it, such as accompanied the remarkable feat of landing man on the moon, will quickly fade and the achievement become just another bit of history — past. Even when he lands on other planets the only benefit for humanity will be a temporary titillation of the imagination and a plethora of materialistic information serving no real purpose whatever. And the scientists will despair at the fickleness of public support and interest, failing to see that while the fundamental aim is not right, mankind is not going to be interested; and that the results are not worth the effort involved, except in arid intellectual satisfaction and in the increased ability to make war, profits and comforts for those who can afford them.

Nothing endures or can endure in the public mind because the foundation, direction and priorities of nearly all of man's organised efforts are self-serving. Only the fundamental needs of all mankind matter — nothing else. Nothing meaningful can be discovered in space while there is so much misery on earth still waiting to be discovered.

Eleven

THE DRACONIC
TRANSVERSE

*Man's evolution is controlled
through a grand design centred
on the constellation of Draco*

THE DRACONIC TRANSVERSE

THE SOLAR SYSTEM is part of the myth of the northern constellation, Draco, the dragon or serpent. In ancient mythology the constellation is associated with death and salvation: the serpent is said to swallow up all souls that at death have not attained to gnosis (real knowledge) and to return them through its tail back to the world where they once again start a new life of struggle towards the gnosis which finally saves them from recurrence. That is just about as far as the ancient accounts of Draco take us. The whole myth, if it has ever been told, has not been recorded with the detailed context and significance I am now going to reveal.

First, let me explain the character of myth. It is impossible to describe spirit as spirit, for spirit must communicate direct to the individual; but through myth it can be done approximately. Myth is the only means at our disposal to describe the reality behind human existence. It is the language, the handwriting, of the spirit, conveying the significance of things perceived; the otherwise untellable truth behind the fragmented physical

173

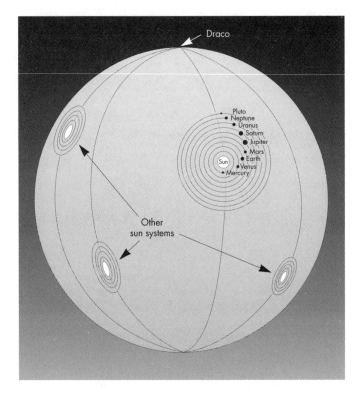

Draco

Pluto
Neptune
Uranus
Saturn
Jupiter
Mars
Sun
Earth
Venus
Mercury

Other
sun systems

The Draconic Transverse

appearance of things. So in reading what follows it is necessary to 'look through' the actual appearance of the part of the heavens I will be describing and to try to relate to the supporting structure of myth or truth behind it. This is what the ancients managed to do. They did not invent the myths. The myth, or truth, was already there, as it is at this moment. It would be completely wrong to regard the Draconic Myth as imaginative fiction. This myth — which necessarily will be seen to touch upon myths of all ages, in all places — is nearer to the whole truth, *the* truth,

174

than any statement that can be made from sense-perceived experience.

The solar system is part of the myth I call the Draconic Transverse. As every myth must have its physical symbol, its appearance in the world, the Draconic Transverse is represented by a gigantic sphere in space, thousands of cubic light years in volume, with the constellation Draco at its north pole. The physical extent of the Draconic Transverse is so enormous that its main stars as seen from the earth are regarded as fixed — never-moving.

As far as the tiny earth within it is concerned, the Draconic Transverse represents our field of reality, the world of myth. We can only observe and know the rest of the universe through it. The rest of the universe can have little reality for us until we at least have begun to understand the draconic mystery or truth which, as I shall show, is so intimate to our earthly existence. The Draconic Transverse determines how we exist, evolve, perceive, live and die to exist again. At this stage of evolution we cannot go beyond it except as it, our reality, permits.

YANG AND YIN

Draco is indeed a dragon or serpent. Its head is Yang and its tail is Yin. The profound cosmic principle of Yang and Yin is the mythical bridge between outer and inner. Yang is behind the apparent external created by our looking out consciously into the unconscious streaming down on to the surface of the earth; Yin is behind the unconscious within. Together, they represent the extent of the unconscious reality, which can only be sensed through myth. Myth alone transcends all conscious parameters.

Yang, the serpent's head, is 'all eyes' — a huge platform of celestial perception. From this vantage point in deepest space, a score of light years beyond the solar system, the Yang principle of Draco presides over time and events which on earth are represented by life and death — the dynamic behind terrestrial evolution. From the earth's viewpoint Draco's tail — Yin — ends deep in the unconscious of the human mind, beyond the psyche, within the mind of the earth, the mind of the one earth spirit. Between the head in space and the tail within is the serpent's body. This is man and all the species both living and dead.

Yang is responsible for the earth's evolution as an intrinsic part of the whole evolving universe. From its towering Draco position it surveys the universe and keeps Yin informed. Yin is responsible for seeing that the correlating evolutionary changes occur within the terrestrial mind. For humanity (as distinct from other life-forms which may inhabit other planets or star systems within the Draconic Transverse) Yang surveys the universe relevant to sense-perception. Also, seeing far beyond sense-perceptive evolution to galactic factors that even our creator Sun is unable to perceive, Yang conveys to Yin the state of the galaxy as it affects the terrestrial mind in considerably more profound ways than we can ever be aware of.

The character of Yang and Yin is will. Yang as Draco's head is the will in the universe; Yin as the tail is the will in terrestrial mind. However, Yang and Yin are not just a closed-circuit Draco to earth polarity. The Yang/Yin earth connection is only one of many comparable meridians linking numerous suns and their orbital matter to form a gigantic cosmic spiritual field — the Draconic Transverse. Within this vast evolutionary system the outer must be kept in unison and harmony with the inner. Neither must get ahead, nor behind. Yang and Yin are the principle — the will in matter and mind — ensuring this.

Yin in terrestrial mind contains all possibilities behind appearances, all that can ever exist in the present and future — all waiting for the moment, the exact time to manifest in consciousness at Yin's behest and precisely to Yang's perfect timing. For Yang is the master of timing in sensory appearance. Yin, deep in the terrestrial mind, is beyond time; it is the principle of abstract knowledge attuned to the original spiritual idea of the earth. Yang determines the time at which any aspect of that idea within Yin's knowledge is released to rise and eventually make its first appearance in the human psyche in the form of differentiated ideas and finally as the objects of the earth around us.

REAL ENERGY

Yang is a principle, not a function. Although it is the positive principle behind the idea of action, it does not act. Yin, the negative principle, is all knowledge of the earth-idea and in lacking the need to know any more, it too is powerless to act. The action connecting Yang and Yin is provided by a cosmic power called Real Energy. Real Energy is not energy as we think we know or understand it. This is because it operates outside our familiar sense-perceptive time-frame. Real Energy is timeless, instantaneous and manages to perform what is impossible in our physical time-gradient: action without reaction, action without consideration, action without pause — instantaneous flow or being. Real Energy is the power which keeps life on earth moving and therefore evolving. Without it life on earth would be static, stationary, impossible.

Real Energy keeps life on earth moving and evolving in the following way. Passing from Yang into Yin in the terrestrial mind it picks up the knowledge of the earth-idea at Level Seven. It 'descends' down through the other levels to sense-perceptive existence at Level One. The energy is thus the conveyor of the knowledge of the earth-idea to the senses and the catalyst of action at the sensory level. The result is the natural world we see around us. It is the combination of the timeless Real Energy and its power of action in expressing the earth-idea through sense that creates being. There has to be being to perceive existence. Without being there is no existence. And the being perceiving it all is the being/person reading these words.

Real Energy acts without creating reaction whereas in our physical world all action produces reaction. Energy expended here is equal to the resistance opposing it, so no real work is ever done. By using force as energy man imagines he creates and destroys while really he creates and destroys nothing. From the exploding of a hydrogen bomb to the tapping of a keyboard, the effects of our efforts are purely formal or cosmetic. Nothing permanent is accomplished apart from the continuation of the human race and the world as an historical whole up to this moment. In our sense-perceptive world of cause and effect, life or knowledge loses its instantaneity and becomes historicity or the past. Because Real Energy is timeless there is no interval in it. But time as we know it consists of interval: everything takes time. Even the movement of the second hand on a clock takes time. And so we live in the past.

What we call energy and use or see acting on earth as heat, sunlight, electricity and nuclear power is reactivated energy, second-hand energy. It works as force, and there is no being, no life, in force. This reactivated energy has had 99.9 per cent of the original knowledge extracted from it. In fact a definition of force is 'energy devoid of knowledge'; for force can only convey information. And knowledge is not information.

At this stage of our evolution it is difficult to comprehend that the main quality of Real Energy is knowledge, because we operate on information. Although we say we have a knowledge of something, we really mean we have experience (information about past activity) or information about something to hand or in our memory. There is no knowledge in this existence.

Knowledge is distinguishable by the fact that it involves no time. Information, on the other hand, always takes time. It takes time for us to receive information, to consider or think about it, speak or write it. We are informationally bound. Without it we know nothing — until we evolve to having knowledge.

To have knowledge is also to know nothing, with a great difference. Knowledge, being Real Energy is timeless. The intervals required for consideration and transmission are absent.

This means that the knowledge being received from Yang is already 'known' by Yin within. The 'transmission' is absolutely immediate. And because the profound knowledge of the cosmic power works for the good of the entire universe, as well as the earth, the knowledge is always: 'All is good' or 'All is right'.

In human consciousness such knowledge eliminates the normal anxieties which can be aroused by information as bad news. The good news in information is only temporary because other information, probably contradicting the good news, is on the way. This is a world of conflicting information, conflicting good and bad news. This phenomenon is typically human and unknown to the rest of the universe where, whatever happens, there is the timeless knowledge: 'All is for the good of the whole'.

COSMIC PARTICIPATION

The objective of evolution can be said to be to rid the world of subjective considerations, which means ridding ourselves of self. When I say that Real Energy and knowledge are 'objective' I mean totally objective, that is to say the subjective (what we regard as time and ourselves) is utterly eliminated from the perception. The result is being, now.

Evolution on this planet, being controlled by the Real Energy acting between Yang and Yin, is universally inspired and is not the isolated phenomenon we tend to think it is. Evolution is a cosmic movement towards cosmic consciousness, towards man's conscious participation in the universe. To participate we must have something real and of value to contribute. In evolutionary terms, our contribution to the whole can be gauged by the condition of our world in relation to the harmony, justice and well-being of all mankind. Our appalling lack of consciousness in these respects will be known to any developed cosmic intelligences in the immediate universe. And any that are undeveloped cannot know about us any more than we can know about them. So our apparent isolation will continue until we have something better to give than our curiosity and selfishness — until we are bent on space-travel not for what we can get out of it, but for what we can give.

It is not the evolution of life on earth that is the point, but the evolution of our planet as a cosmic organism, together with the other planets of the solar system and all other cosmic bodies and stars within the Draconic evolutionary system. This earth-life that goes on and on yet seems to end perfunctorily and witlessly in death for the individual, is not your life or my life but the life of the planet earth evolving towards a greater consciousness. Both in living and in life after death we all combine like cells to form one single evolving terrestrial being within the universal reality. Through the Yang-Yin Real Energy connections of the Draconic system, our sun, the earth, the planets and other stars and cosmic bodies contained in it communicate with and inform each other, exchanging and integrating their continuously changing evolutionary tones or states.

We have nothing to contribute to cosmic evolution except our death. Man, together with all the species, lives, dies and recurs in the one-way incoming energy system of the Draconic Transverse where the energy that is continuously being received into the three planes of human existence forms man's individual body, nature, character, emotional and mental responses, circumstances and world. But no Real Energy whatever goes out from him into the cosmos until he dies. This energy, the value of his life just lived, becomes a part of cosmic evolution by passing through the death band into Level Four. There, Yin — Draco's tail within the terrestrial mind — absorbs the information and passes it into the full Draconic system of Yin and Yang which unifies inner and outer existence. At our present level of development only life and death — the Draconic syndrome — produces real or cosmic spiritual energy.

The species as a whole, and man in particular, are the chemical agency of the earth for receiving and converting the cosmic energy to the frequency of terrestrial mind. Man's life is ingestion, his living experience conversion, and his death process distillation. He converts the energy to his unique individual frequency by the pleasure/pain of his life experience; and on death, this passes into the Draconic system.

Astrology

As the real value of a man's past life passes out of terrestrial mind via the Yin/Yang Draconic connection, it disperses and manifests throughout the entire universe. Because the man has always been a part of the universe his energy pervades even the most distant nebulae and is in every energetic expression originating anywhere in the universe. Every cosmic influence received on earth has to have his number on it.

For these cosmic influences to be received on earth they first have to be converted to our particular terrestrial frequency. This is performed by the sun's radiation. Acting like a massive broom or an electro-magnetic brush ceaselessly sweeping the vast area prescribed by the orbiting planets, it picks up the influences from these planets as well as from outer cosmic sources entering the solar system. These it monitors, down-grading and up-grading them to the suitable frequency for life on earth. The information is transmitted in photons or light particles to the mind through the eyes and other tissues.

This process is the basis of astrology — man's most successful attempt so far to formulate the mythic principle behind the Draconic evolutionary system. But it must be understood that astrology is only a partial approximation of the truth. Due to psychic distortion in the human mind and insufficiently perceived data, it falls short in its application to specific phenomena.

The new life a man makes for himself in the death world passes as Real Energy through Yin into Yang to be added to Man diffused throughout the universe. Showering down onto the earth from the rest of the cosmos, the new life is absorbed by his mother-to-be through sunlight as well as through the food she eats and the air she breathes. These influences — released as time by Yang — are the external factor that at the right moment will trigger the man's conception and birth, just as eventually they will trigger his death. At the precise moment of his birth (or conception) the entire universe, and more specifically the sun and planets, will be in certain positions relative to the earth. Together these configurations — correlating to the place, year, month, day and time the man is born — will present the nature of his self-made future in the language of universal symbolism. If the positions of the cosmic bodies can be mapped and the symbols correctly interpreted for the moment of his birth (the earnest endeavour of astrologers down through the ages), the result will indicate not only his self-made nature or future but also the cosmic assistance and resistance he will encounter in endeavouring to realise the full potential of his new life. Furthermore, the subsequent positions of the cosmic bodies will show the events affecting the man's life at any particular time.

MAN'S SENSE OF TIME

Man is a composite of three notions of time on which the cosmic forces act — the future, the past and the present.

A man's future is what he constructs after death out of his power to love. This is his real future — not his imagined or projected future or his bid to escape from death. Because it is real it has no interval, no sequence, and therefore cannot be conceptualised or thought about. This future consists of two aspects. As Yin it is the abstract outline of his next life left in the Yellow Plane at Level Three. As Yang it is represented by the sun and other cosmic bodies that trigger the new life experience. The man's self-made future will be the broad outline that his life will follow and, as his inner Yin, it will assert itself as his feeling of free will. But in fact it can never be lived as he intended or designed it. He will always yearn inwardly for something that never eventuates to his satisfaction. This is because the future he has made for himself will be modified from its inception by external cosmic influences which represent the evolutionary contribution of the rest of the immediate universe to his life. As he affected the universe so it now affects him.

A man's past also has two aspects, the Yin and the Yang. In its Yin aspect the past is in man's mind. This past is a chemical self that originates in the Red Plane and forms an emotional body around the abstract outline of the man's future. This body is the material man whose involuntary self-centred motivation is to survive and reproduce itself as an emotional and organic vehicle for perpetuating the species. It is this 'old' man in the man which will often confuse him in his new life by seeming to work against his self-made future. The Yang aspect of man's past is represented by an external cosmic body — the moon. Acting on the fluids of the body the moon's influence strengthens the man's emotional identification with the past of fixed perceptions, memory impressions and family, social, national, religious and cultural traditions.

The interplay in man's consciousness between his future and past creates his awareness of the present. This is where beauty and the beast, the agony and the ecstasy, often seem to share the same moment in time within him. In the mind, the present is a swiftly oscillating point between the Red Plane and the Yellow. This point varies in everyone: the less evolved man living in the emotional, material now and the more evolved man living more in the moment. However, in every man's life this awareness of the present — acting sometimes as one aspect and sometimes as the other — will very often seem to be working against both his future and his past.

Remember that in most people for most of the time the Yellow Plane is almost constantly being invaded by insurgent Red Plane self and consequently the intellect in most people is coloured by emotional interference. While a man is performing work on earth or being as objective as possible (dealing only in facts) the power of his intellect is relatively clear. But the moment he thinks about the past or acts from his emotional attachments (which are as numerous as his thoughts about the past) the Yellow Plane is invaded and he is once again subjective and unreliable. For humanity every thought and consideration is in the past; most of the time is spent thinking or talking about the past and to a lesser degree about an imaginary future. Very little of the world is ever perceived as it is in the present moment; it is seen through the screen of past impressions and opinions, likes and dislikes, regurgitated feelings and false sentiments.

THE INFLUENCE OF THE SUN AND MOON

For the most part people live out of the past in the world of the senses. Invasion of the intellect by the self is felt in its most noticeable form (if felt at all) as changes of mood (Yin) manifesting as changes of behaviour (Yang). But more widely, behavioural changes represent cosmic and planetary influences transmitted by the sun (Yang) acting on the emotional self (Yin) via the brain and glands. As the sun, symbolising enlightenment, actually illuminates the moon, symbolising resistance and attachment to the past, so it forces changes on the sentimental and emotional self. As the moon's influence holds the man back by making him resist change within himself, the sun's influence goads him on to dare to dare and fulfil his potential. The moon makes him seek the safety of the herd, or the comfort of the womb, while the sun tries to tip him out into the world to live creatively, originally and individually in line with the life he scripted for himself. And gradually, through the traumas of living, man's lunar attachment to his emotional organism is broken and converted into solar love — cosmic consciousness.

Twelve

THE CREATION OF
THE UNIVERSE

*What is the real answer to the biggest
question facing science?*

THE CREATION OF
THE UNIVERSE

THE TWO MAIN scientific theories of the universe are the Big Bang theory and the Steady State theory.

The Steady State theory, currently out of favour, is that the universe is infinitely old, or has no beginning in time; that it has always fundamentally been the same and always will be the same, and although it keeps expanding as the galaxies race apart from each other, new matter is being continuously created in the space between to make up for the decrease in density; with the result that the universe will always look the same from any point in space or time.

The dominant theory is that the universe began with a Big Bang. An infinite amount of space and matter, it is believed, was compressed into a tiny finite volume which exploded and has been flying apart as the universe ever since. Deficiencies in this theory are recognised by science and stem mainly from the fact

187

that the calculations involved go back only to what scientists consider the first split second after the Big Bang. The instant before their calculations — the state or origin in which time equalled exactly zero — remains a mystery.

What happened before the Big Bang? That is the awkward question posed by the theory. And if the Big Bang was the beginning of time, as some scientists apparently are starting to wonder, what was, or is, the pre-time state?

In the following I answer these questions. I describe the state in which the physical earth and the solar system came into existence. In doing so I necessarily refer to the rest of the manifested universe and later I will have more to say about the character of space, time, matter, life and intelligence. Furthermore, what follows shows that both scientific theories of the universe are correct in their own way; and that it takes both views — or an apparent contradiction — to adequately describe the origins of the universe.

THE REAL BIG BANG

The pre-existent state behind the Big Bang and all existence is infinite mind — an inconceivable void empty of both space and matter. Implicit in infinite mind is infinite will.

The creation of the universe began when infinite mind had the idea of existence. I cannot say where this idea came from as I can in the case of earth-ideas and solar-ideas, for that would suggest infinite mind was not infinite in its own right; perhaps something is more infinite than infinite mind and is beyond any description. The instant the idea of existence occurred in infinite mind, the will separated from the mind and acted on the idea. It encircled a huge area of infinite mind making a girdle around the potential space of the universe inside. The sudden separation of will and mind produced the most tremendous retort, the biggest bang of all time. This continues to reverberate in the mind and is behind the rationale of the scientific Big Bang. To exist, all things — even erroneous theories — must take their rise from original truth.

The will, which can be described as the executive action of mind, carried with it the complete idea of existence — all that could ever happen. This was retained as the outer side of the girdle facing infinite mind. The inner side facing existence consisted purely of the enormous power of will. The outer side formed the intellect.

It is possible for us to follow the formation of the universe as an event in our own experience. But we have to appreciate the two distinctly different points of view involved — the mind's position and the existential sensory position. Both are familiar to us all.

The mind's position derives from the infinite mind in which the universe manifested. This still, pure state containing nothing — and yet subsequently everything — is involuntarily and approximately participated in by man in dreamless sleep, in unconsciousness and sometimes in meditation. A version of it can overtake people during their waking hours, producing an extraordinary feeling of wideness of vision or being. In this state, the whole natural world can seem to exist within the observer's awareness and to have a very limited or relative verity by comparison. The still-mind position is not normal for man at this stage of our evolution. The sensory position is the normal viewpoint.

We can now gain some understanding of the formation of the universe by shifting between the mind and sensory viewpoints. First, let us take the mind viewpoint. Let us imagine that our mind is the mind which is about to contain the universe . . .

All is still, supremely at rest and at peace. Suddenly there is a tremendous bang. Immediately we are aware of something, literally jolted into knowing something. We do not know what it is except that our consciousness which unknowingly had been infinitely wide and uninterrupted, is now cut off. Something has come into the mind, our mind. Where there was nothing, there is now something. It is a mind-riveting, mind-limiting event.

A sensory reflection of that something can be observed at this moment in our own minds. It is the grainy blackness, the dark lanes and pinpoints of light in front of our awareness when we close our eyes, or have them open in complete darkness. This grainy screen, closing us off from our pristine consciousness, is the intellect, the energetic reservoir of all the ideas of existence. It is present in all creatures at all times, even in the blind.

The birth of intelligence

The intellect is the reflective screen that allows us to perceive and reason. Its formation also marked the beginning — on its inner side — of the external, sensory universe: of time, the solar system, and the potential for intelligent life to develop there. As physical beings we are provided with the sense of existence by the stars, space and matter; while as mental beings the intellect allows us to evaluate that experience as information and to evolve rationally.

So we have our eyes closed, are observing in the mind the grainy black intellect in front of our awareness, and then we open them. We are immediately sensory. At the speed of an opening eye-lid we have moved from the inner to the outer position. A second ago we were observing the intellect — not knowing anything (unless we were thinking) — and now we are sensory intelligence in our own evolutionary earth-world.

We can see, dig into and feel the earth, the planet that is ours in the solar system, and we can look up and see the universe of stars. What is it we are using to know this — and which was not needed a second ago, when we were simulating the stillness of pure mind? Intelligence. We need intelligence to be able to know. We do not need intelligence to know nothing, to be the still mind. In fact, intelligence is what we are: various degrees of it, representing a gradient called life on earth. We are the intelligence that has evolved in time out of the space and matter of the solar system, and specifically the earth, on this, the universal side of the intellect.

Through the intellect — and I mean literally by passing through it — we have emerged to become intelligent life on earth; just as by closing our eyes and being still we can return back through the intellect to the pre-existent state. The only proviso is we must not want, or think about, sensory existence. To do that automatically transports us at the same incredible speed back through the intellect into sensory existence on earth — with all its dynamically needling evolutionary traumas and deliciously addictive experience.

Intelligence, for all its activity, is the spatial or sensory equivalent of the original still mind. To this state of stillness it must eventually return by evolving through experience of the universe.

THE STATIONARY UNIVERSE

According to science, the earth is moving through space at fantastic speed. It is said to move into new space around the sun at 30 kilometres per second, forward with the sun at 20 kilometres per second and around the galaxy at 220 kilometres per second, all at the same time. But on a cosmic scale as great as the Draconic Transverse — bearing in mind that reality is totality and the greater the totality observed the more the reality — the earth is motionless. In fact, the motion of all heavenly bodies is due to the oscillation of our observing intelligence and not to any real movement of those bodies.

From the point of view of mind — the original still reality — the universe is motionless and has been since the instant it manifested as the sensory representation of the intellect. Only on the sensory side of the intellect is there movement or evolution of objects in time. The intellect has never moved, never been added to or detracted from since the Big Bang that accompanied its emergence. When we observe it in our heads it is the same as it always has been and always will be. Any apparent movement or change is due to the unsteady state of our evolving intelligence as spatial beings.

The intellect is the stationary bridge between the universe and infinite mind. Intelligence, evolving out of matter and life on the universe side of the intellect, eventually crosses it to return to the timeless and motionless original pre-existent state of mind. It is intelligence, not the mind, which is evolving in time and space. Intelligence advances towards becoming pure, still mind by its understanding of change and motion through universal experience. Intelligence, in this respect, can be described as an intellectual or reflective movement towards the understanding of a reality greater than that suggested by the senses. Intelligence acts to know, to sort, to uncover. By reaching out through the senses and reflecting its experience off the intellect, intelligence comes to know physical existence and to give it significance and value.

Intelligence, wherever it appears in the universe, oscillates at different frequencies. The swifter the oscillation, the higher the intelligence; motion slows down significantly and the perceiver sees through existence. From the earthly point of view, the frequency of intelligence varies with the matter (the person or species) it arises in, as well as with the relative reality of the symbol or object being observed.

For instance, if a person can keep his attention intelligently focused on the intellect — that dark screen of minimal experience in front of his awareness — he will eventually lose all sense of time and motion and enter the infinitude of original still mind whence he, time, motion and all objects originate. He will remain in this state until his spatial (physical and emotional) self starts to exert its restlessness (evolutionary need of motion) and demands his intelligent participation in worldly affairs. Intelligence, in spite of its ability to detach itself and focus on the intellect, is bound by time to the matter or body it arises in, and must complete that particular evolutionary draconic life/death cycle.

All motion on the earth is due to the relatively slow oscillation of the observing intelligence — that is, our sense-perceptive condition — as well as to the relatively superficial significance we normally attach to it. Not much reality is apprehended in our lives; thus we are surrounded by movement and restlessness. When reality is faced or comprehended — an experience often associated with death or loss — there is a distinct slowing down both sensorily and intellectually; a greater significance or inevitability is seen.

In our physical or sense-perceptive experience, the stars represent the highest reality. The greater the reality the faster the observing intelligence has to oscillate and the slower the star is seen to move. Similarly, the further a man is from perceiving reality in mind or sense, the slower his intelligence oscillates and the swifter and more numerous are the objects and thoughts that shoot through his mind or space to distract him. For example, an unintelligent person would not be expected to study the cosmic reality of the heavens, or his own mind. If he did glance towards either, instead of reality he would only see the various intellectual inventions he had formulated of it, or that had been formulated for him. These concepts, being unreal, would move through his mind at great speed; and, being unable to slow them down or stop them with his limited intelligence, he could be said to lack power of attention, or depth. Such people become easily bored and desire continuous movement and stimulation — being addicted to the need for the constant passing of time in the form of objects or thoughts in their minds or in the space around them. For an intelligent observer, time, like the imperceptible motion of objects in deep stellar space, is almost non-existent. The reality a man can observe at any time is determined by the stillness of his mind or his space, and this is a measure of his intelligence.

EVOLVING INTELLIGENCE

Intelligence evolving out of matter and life eventually exceeds its life-form but without necessarily leaving it. It does this by speeding up, quickening. This is accomplished very slowly through evolution so that finally intelligence reaches a speed equivalent to a time-change that enables it to perceive aspects of other time-gradients in addition to its own. The speed at which intelligence oscillates determines the character of the time.

The human life-form — the organic body of humanity and the species — is fixed. But although anchored to the human body, or the earth, human intelligence can still exceed the sense-perceptive time barrier and go beyond its own gradient. (Endeavour to think of human intelligence as a single gradient, as intelligence on earth, not as numerous individuals more or less intelligent than each other.) By quickening intelligence towards the top end of its gradient, evolution demands that intelligence must free itself from the ingrained habit of identifying with the particular life-form or time-gradient it occupies. At the present point of evolution it is somewhat rare for human intelligence to realise that it is separate from the body out of which it emerged. But the realisation of this does happen and is identical with a time-change: automatically other life-states or worlds are then perceived direct without sense-perception and are known objectively — that is when the subject, the self has been exceeded. This is common in mystical experience.

Because human intelligence has not yet evolved sufficiently for us to spend all our time on the mind side of the intellect, we have invented death (or it was invented for us) which is a sort of in-between world. The world of death, or the living dead, together with the world of the dying living, make up a wondrous whole of draconic gnosis, the world of Real Energy or objective knowledge.

Draconic gnosis is a state of evolved intelligence which life on earth has achieved at the narrow, higher end of its time-gradient. It includes the mythical, occult and magical worlds. Draconic gnosis is not pure mind for in pure mind nothing, not even gnosis, can exist. But it is closer to purity or stillness of mind than has yet been consciously achieved by most of humanity as the intelligence of planet earth.

The world of motion

As we proceed we must keep in mind the two sides to our existence as earthlings: the side of infinite mind which is the real side we always retain, even in sleep and unconsciousness; and the actual, external, sense-perceptive or transient side which disappears in sleep and unconsciousness.

Actuality is our familiar sense-perceptive world of motion and interval. And it is entirely ratio-retrospective, meaning that to know its significance we have always to think or calculate back across an interval to a supposed beginning. The naming of things is one way we have invented to do this quickly. Even to recognise a person or thing we involuntarily have to refer back. To make sense of anything means to go back. Our perception or particular time-gradient demands a beginning for everything. So it is inevitable that we postulate a beginning of the universe.

However, it is impossible to calculate back to the beginning of anything, because every calculation implies a pre-existent motion or time. You can calculate back to when a bullet leaves a gun but that proves nothing about the beginning of the bullet or the gun — only that motion or time never ends so never begins.

You can calculate back to the moment of your birth. But that signifies nothing about the beginning of you. If you go back further and calculate the moment of conception you are no closer to the start of you: the only answer will be in terms of motion or time. You cannot calculate back to the beginning of you, or the universe, because you always arrive at motion and motion is not the beginning.

The original motion happened in reality when the intellect manifested in the mind as I have described. In reality, this is the only event there has been or ever will be. Whether perceived or not, the intellect is there: it is always there and always has been there. For this reason, the intellect cannot be said to have ever occurred: the question 'When?' arises from the incompleteness of its shadow, evolving intelligence, looking for a beginning or end to itself. The intellect for us is the primary state, the ever-present motionless reality that makes possible the fact of intelligence and therefore the actuality or external world.

To summarise: nothing moves in the mind, in reality. But everything is in motion in the actual world. Reality consists of mind and intellect, both of which are absolutely still; and actuality, the ever-moving external world, consists of evolving intelligence whose restless urge-to-know is the cause of all motion.

PERCEIVING REALITY

Science's Big Bang theory logically postulates the beginning of motion. But there can be no beginning to motion because motion is not real. This the theory itself demonstrates by completely collapsing — it fails to be able to go back any further — at the precise point where the beginning of motion and the universe is expected to be found.

Motion (and the Big Bang theory) is an effect, a condition created by developing intelligence. By trying to discover the beginning of motion we are trying to discover the beginning of our own ignorance or unintelligence: an impossibility. It is like looking in a mirror and trying to see who's looking; or trying to lift yourself off the ground by your own shoe laces. Motion, as I have said, shows that the intelligence observing it has not evolved sufficiently to see the truth that reality is stationary.

I am not attempting to deny the actuality of our world. Our world is created as we all know by sense-perception. What we do not know is that the sense-perception is of an evolving intelligence. And it is that intelligence, the perception behind the senses, that creates the movement of things and this passing world of ours. That intelligence — we might call it terrestrial intelligence in the absence of any other planetary life seeming to share in it — is evolving through life on earth. Humanity as a whole, and individually, is that intelligence. That is why we all see the form of the world and the universe the same as each other.

What I am proposing, and will be proposing in different ways throughout this book is that terrestrial intelligence has now evolved sufficiently for some of its individual cells to perceive this new truth that the reality behind the senses is a stationary world of infinitely greater power and significance for man. The only intelligence that can perceive the stationary real world or universe behind the moving actuality is that which admits of no past in that moment. We all have access to this 'top end' of terrestrial intelligence because it is our planetary intelligence — if only we can surrender reliance on the past or interval. It operates now, in timelessness, undistracted by the motion of things, the doubts and confusions of the past which are the waverings of the lower end of our intelligence. It surrenders the past anew every moment — memory reference, attitudes and all dependence on information outside direct perception now and in fact, dies to motion. As a result it perceives reality direct; that is, without the pause or interval that is sense-perception.

LOOKING INTO ETERNITY

The Big Bang in the mind, the advent of the will/intellect, is the first cause of all motion or evolving intelligence. But the mind cannot know this until intelligence evolving out of that motion in the actual world — you — discovers that truth, as we are endeavouring to do now.

As we have already seen, the advent of the will/intellect, that one and only real event of all time, unbeknown to the mind, simultaneously created the universe, the obverse side of the intellect. Like a coin, on one side was the intellect, on the other the universe, each unbeknown to the other. But as both sides of a coin cannot be seen together, it would be impossible to perceive or write this description — except for the truth. The truth, that most wonderful but elusive quality, lies here between the two sides, separating and yet joining them as the moment of eternity.

Eternity is neither one side of the 'coin' nor the other but contains the truth of both. It is the supreme abstraction, analogous to reducing the thickness of the coin so that finally it is both sides at the one time. This, eternity manages to do. It manages it by preserving forever the moment the intellect appeared in the mind. As that moment never ends it never had a beginning. This absolute timelessness of eternity is a discontinuous, motion-less, energetic representation of all that ever can be or will be enacted in the slower pure time of Real Energy and the infinitely slower light-speed interval of our sense-perception.

Real Energy is the executive power of eternity and to our perception an inseparable part of it. It is the means by which eternity is attenuated, drawn out, into a replica-in-motion of eternity — our actual, sense-perceived world.

I have just described eternity as an energetic representation of all that ever has happened or can happen in actuality. That means it is perceivable, but how? It is perceived as the now. The moment of eternity, the moment of truth, is now.

Now is the one and only perpetual instant from which the world continuously — or rather discontinuously — begins afresh every moment by having no past.

THE ORIGINAL MOMENT

The beginning of the universe and man is not some time in the past. It is now, this moment, which is demonstrably the same now as always. What stops us from realising this is that we regard what happens as the only important consideration and dismiss entirely the continuous moment it all happens in. If we could give equal cognisance to both we would immediately perceive the truth of now, man and the universe. But as it is, we are distracted from the truth by perceived events which immediately become for us fixations of the past. Each of us measures his life by those events, and so we live mostly in that past.

Actually, all we are doing with the years of our lives is measuring the past, the interval we have left behind, our distance from our assumed beginning, our birthdate. Science tries to do the same thing with the Big Bang theory by giving a date and start to the universe. But both beginnings are illusory. Reality is neither past nor eventual; it is now. Now is the original and only state of things. Nothing ever has happened or can happen outside of now. Although self-evident, the significance of this is extremely difficult to concede. We are involuntarily, unconsciously terrified of it for it means giving up the past and seems to predicate losing all of the established foundations from which we gain our psychological sense of security. Through the feeling of having been, we get the feeling of being someone, something. It is a substitute for being now. It is comforting, reassuring almost to the point of necessity for most of us; but it is not the truth. To be someone or something at any time requires living in the past.

To be able to perceive the beginning of the universe and oneself as now is the beginning of immortality in the individual. To cling to any other beginning in the face of the real is for the individual to continue to calculate his own inevitable and equally unreal death.

Certainly, each of us was born as a life-form on earth. But that proves nothing apart from the fact that each of us was born as a body. To regard the birth of any form as a beginning is totally presumptuous. The question is, can the reader recall any time when he was not? The answer is no. But if I cannot remember, it is an absurdity to assume that there has been a time when I was not what I am now. This is the assumption of death of which most of us are guilty. Our delusions change as our past grows, but we do not. No man feels he is 40, 50, 60, 70, or any age. He may feel pain and the restriction of the body — his past — but the man, the pastless I in him, never feels any older. However old, the man is ageless. I am not an event: not a body, a birth certificate, a memory or a bathroom mirror which by reflection on the past that is my body, reveals to it its age, past or mortality. That body and all that helps to measure it by reflection and

199

memory will die. But I cannot — any more than my birth can be my beginning.

To summarise: we are beings of now, as the universe is the expression of now. As part of the universe we cannot be separate from it, except through the illusion of past created by our evolving earth-intelligence. In truth there is no time but now. Now is exactly the same instant that the intellect formed in the mind and the universe began. In truth nothing has changed, nothing ever will. The thing that does change and therefore is evolutionary, or at any moment not yet itself, is intelligence. Intelligence invented interval, the past and the Big Bang theory because it is not yet up to living in the now.

I am proposing that it is possible for man to 'jump' his own time of past and interval and to realise for himself that eternal moment in which he and the universe are as they always have been. This, I accept, may need a good deal of demonstrating as such a proposition is contrary to normal experience and reason. However, what is normal and reasonable arises from memory and past experience and is itself part of the successional time trap that has to be vaulted. It is universally true that any real proposition — the positing of even the highest reality — can be demonstrated through the intellect, provided the intelligence (as the evolutionary position) of the observer is up to it. As I will describe later, the same intellect is used by ultra-intelligences throughout the universe as well as by ourselves and the living dead, and is therefore equal to any demands that can be made of it. It manifested in that original first instant which is preserved forever as eternity, and it is therefore outside the successional time trap that we have to vault.

Thirteen

POWER IN
THE UNIVERSE

*What impels the formation of
space and matter?*

POWER IN THE UNIVERSE

THE FIRST THREE PRINCIPLES

I HAVE DESCRIBED the creation of the universe as the instant the will/intellect separated from infinite mind. I will now explain how this happens and how the intellect itself comes into being. But remember that everything I am going to say happens in another time-gradient to ours where there is no past, no interval. All the 'events' are happening simultaneously, now, deep within the psyche, your psyche. The models and sequences I use are therefore only indicative of the essential idea.

Three inseparable first principles are behind the moment-to-moment existence of the universe: infinite mind, infinite being and will.

Infinite mind has to be understood as an unending void of 'nothing'. Implicit in infinite mind is being. This means that being is everywhere in the mind; although when being is realised by an individual he knows it to be the centre as well as the whole. Consequently, being can be regarded as the centre of infinite mind from any position.

203

We know very little about infinite mind and being, but we know a great deal about will. Will is the power in the universe. Like infinite mind and being, the will is imponderable except as effect. The existence of the universe is the work or effect of the will.

The will is equal in primacy to the two other principles. In fact from the created point of view (ours) it may seem to enjoy greater freedom and privilege. Will is the power of complete freedom to effect a cause. Moreover, it can 'travel' through infinite mind and being without affecting either infinitude. But this is conditional on it remaining 'straight' and not starting to curve or curl around itself. This is demonstrated in the spiritual life where the will must be one-pointed (straight) and not curl or curve into a desire pattern which tends to go round in a circle. The will has to be strong enough to overcome desire for other things. As soon as will curves however, there is what I call a prime effect. To us all prime effects are causes. But there are no real causes in our existence, only effects.

THE WILL'S FIRST CREATION

The creation of the universe — the first prime effect — began when the will started to curve in a gigantic circle around a section of infinite mind to form the paradigm of existence. Curvature is the start of time and the beginning of past which is the essence of existence. Infinite mind, with being at its centre, is timeless; it has no past and therefore cannot be in existence. But the will's encircling action threatened to make infinite mind exist as the circle closed around it. For the briefest of moments, before the circle closed, infinite being was exposed as the potential centre of existence. (But this could never be, other than as the instant of man's God-realisation or the realisation of being, both of which occur in the infinite mind, the consciousness outside existence.) Infinite being cannot be contained or encircled by existence; so it vanished, withdrawing at the centre 'back' into infinitude.

Meanwhile the will had formed a girdle of astronomical proportions. Inside the girdle was existence; outside it, infinite mind. To grasp the vastness of the inner area comprising existence, remember that the Draconic Transverse, all it contains and all the other constellations and gallaxies known and not known, are all parts of existence. As far as we are concerned there is no end to existence. Any attempt to go beyond it will always be frustrated by the curvature of the will expressed as the girdle. Whatever makes the attempt will be forced into an ellipse and back into existence. The only way out is psychological or spiritual — to be as nothing; that is, to realise the infinite mind on the other side of the girdle. This is the evolutionary endeavour of all intelligence in existence.

While forming the girdle, the will also revealed the intellect which had been implicit in will. As I have already indicated, on the existence side of the girdle is the sense-perceptive universe of stars, whose creation I will describe shortly, while on the abstract other side — as on the other side of a coin — is the intellect.

Having closed the girdle, will was now facing will across the space of existence. This unique confrontation of will to will generated an ever increasing and incredibly intense build-up of power. Radial lines of power from all round the girdle converged on the centre-point through which infinite mind and being had withdrawn. Power at the centre intensified and started to build up back along the radials towards the implacable will. With no outlet, the concentration of power peaked at the centre where infinite mind and being had vanished, and universal I, or intelligence, started to arise. Finally, under the relentless pressure, universal I burst into full consciousness as the first expressed principle of intelligence behind all cosmic idea, knowledge and existence. Without I, there is nothing.

The encircled area of infinite mind had now become universal mind; what was infinite was now conditioned by having a centre, I. This I is not absolute because it is not infinite, like being, but it

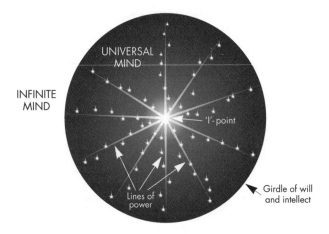

Stars manifesting on lines of power

may reasonably be described as ultimate, the step before what is absolute. So I, as the centre of universal mind am the ultimate truth.

In other words 'I' is the centre of the universe. I am where I am, and that is the centre of the universe, just as what I am experiencing now as the centre of my individual awareness is I. That I centre can be any I: from I the writer or reader of these words to I the spirit of the earth, I the consciousness of the solar mind, I the almighty Lord of the universe. In ascending octaves or gradients of time all are finally united in the one mighty I.

THE FORMATION OF THE STARS

Around the intensifying radials of power, space started to form. In this space (science's positive vacuum) the universe of energy and matter was about to appear. Simultaneously, the power that was building up at the centre I and extending back along the radials reached such intensity that points along the lines crystallised as stars and matter in the newly formed space.

On the infinite mind side of the girdle this last event had the effect of forming all the ideas behind existence and impressing them on the intellect. Whilst on the universal side of the girdle are the stars and their systems, on the infinite mind side are all the pristine ideas of existence held in the intellect. The intellect is completely reflective of all knowledge that intelligence can ever discover and reflect on.

I repeat, all I have just described occurs simultaneously as the one and only real event and moment that has ever been. This moment is the now which also is the moment of eternity. Only in retrospect (thinking back) does anything else ever happen — a subjective condition due purely to the evolution of sense-perceptive intelligence which I am.

Stars are really apertures or openings to the one reality behind the intellect: each star is a different aspect, or window. Stars are power. They are the nearest thing to simultaneous being and non-being as is demonstrated by the paradox of light. All light comes from the stars sometimes acting as a particle (being) and other times as a wave (non-being).

Stars are present in the universe as gravitation. Stars are gravitation. Gravitation is the one constant reality from which all subsequent existence arises. And it is present only in stars although it appears to be in planetary and other orbital matter. Each piece of matter takes its gravitational power from the star or stars whose system it is in. Stars do not consist of matter such as is found in the planets. In a way, they can be called ethereal, non-existent apart from gravitational influence which makes them discernible through their subsequent effects on matter. The universe is a gravitational field of effects provided by the stars, and principally, as far as our earthling intelligence is concerned, by our own star, the sun.

What we perceive, measure and speculate on as the sun is not the sun at all but its effects on matter. The sun is not really there

— all we see are subsequent effects inherent in the causal power it symbolises. We earthlings are eight minutes — the time light takes to travel from the sun — from that nearest solar window to reality. Existentially we can never get much closer to reality than that; our physical vulnerability fixes our position pretty firmly. For the scientists, physical existence offers no end to what can be discovered and understood of the effects of reality — without them actually getting any closer to reality in fact or truth. Reality can only be approached direct, not through its effects. And this we must do by freeing ourselves from the gravity of the past.

THE BIRTH OF THE SOLAR SYSTEM

Planetary matter manifests at the same instant as its star. Our sun did not manifest ahead of the planets. It and the stuff of the planets — the entire contents of the solar system — came into existence together as one complete event. Thus, original matter in the planets, including our moon and meteorites, will always be found to be the same age.

It was the sun manifesting as an effective star on its radial of power which gave reality, or beginning, to the solar system. But as the primary effect and therefore the only event outside of time, the sun could not manifest on its own or ahead of the other primal parts — namely the planets — which were essential to the system and all that would arise from it, including the intelligent life which is ourselves. The entire system had to manifest together, or none of it. Any object appearing later could only be a sub-effect and would not possess the original reality or permanence of that unending first instant. That incredible instant, the moment of perpetual reality beyond successional time as we know it, is preserved forever in the unaltering gravitational power of the solar system — our symbol of eternity.

In relation to the instant when the solar system manifested, nothing has changed. Much has happened in successional time, but nothing has really changed. No new matter (energy) has joined the solar system and none has left. Any apparent incoming or outgoing is due to the action — really the limitation — of our evolving intelligence. This is dependent on interval, signified by the slowness of the speed of light, for the transmission and receipt of information. If all the mass or matter in the solar system could be measured today against the original contents, exactly the same amount would be found to be present. Forms change but mass or energy can neither be added nor lost. Apart from the fluctuating frequency of intelligence, the whole universe (as gravitational power) is stationary.

POWER, GRAVITATION AND FORCE

The existence of the universe depends on three types of energy — power, gravitation and force.

Power is cosmic, the supreme degree of consciousness called spirit. Spirit is the original state of the universal mind before the matter of the universe appeared. As the will that formed existence was the will of infinite mind, so the 'lesser' will of universal mind is spirit. The void of universal mind is a potential of power lines or meridians on which stars explode into existence. Our sun formed on a power line extending from Draco in the north to another constellation in the south.

From the sun emerge gravitation lines. The sun's first gravitation line is represented by the body of the sun and the remainder by the planets and their orbits. Planets, remember, have no intrinsic gravitational power. Gravitation is the property of a star, and intelligence is the ultimate property of the matter orbiting it. The power of gravitation in the star draws life and finally intelligence out of matter. All gravitation is an extension of the gravitational power of the central sun or suns in any system. Planetary matter merely serves to show the Real Energy points at which life is possible within a star's gravitational field. Where matter is, there is the potential in time of life and intelligence. No matter can exist outside the influence of a star.

Force lines are the third set of energy lines. Matter appearing on gravitation lines immediately polarises into force lines which emerge from each of the planets to form fields of planetary intelligence. For us the force lines of the planet earth represent the range of our human intelligence.

Power lines, gravitation lines and force lines each have their own time-gradient or speed of time. As power lines represent the world of spirit, their time is beyond our comprehension. Gravitation lines represent the world of Real Energy; and force lines in our case represent our physical, sense-perceptive world of past and interval.

Force lines of the planet earth consist of information travelling at the speed of light. Energy travelling at the speed of light cannot be instantaneous as Real Energy is; it is always 'behind' the moment of origin and thus forms an interval of elapsed time or past. It is due to the slowness of our force-line intelligence that light from the sun, the nearest star from which we receive information, is calculated to take eight minutes to journey to earth. Similarly, light from the next nearest star is four years old by the time it reaches us. That star could have exploded and vanished yesterday and we would not know about it for four years. Even after the star had ceased to exist we would still be seeing it as it no longer was for another four years. We can never know sense-perceptively what is really happening out there in space, only what has already happened. While we cling to the limitation of force-line thinking — intellectual materialism — we are compelled to live forever in the past.

Of course, in the intimately closed-circuit system of earth existence it works reasonably well. But associated with that same reasonable physical or sense-perceptive time-gradient is the dreadful personal limitation of death and mortality. Here, and in this book, we are endeavouring to understand and so enter the swifter time-gradient in which interval or death is exceeded because that pause or past, caused by the delay of information, is eliminated. In that swifter time-gradient knowledge is identical with what results from it, whereas to our perception information always precedes or follows the result. This means that in the swifter perception there is no past, no interval: everything is new and as it is every moment without any previous moment (past) to give it continuity.

There is only one thing which moves at this faster-than-light speed, which continues without continuity, without past and that is life. Life is gravitational. Living, which is our conception of life, moves at a snail's pace beside it. Life, being at the same time knowledge and its result, the beginning and the end, knows no interval and hence no death. Gravitational consciousness which observes life in this way therefore marks perception of the state of immortality.

When individual consciousness is grounded in gravitation lines no interval is available for any conceptual existence, such as our selfconsciousness occurs in. Such a self-conscious or self-terminating existence as ours is perceived to be unreal against the moment-to-moment presence of all things without beginning or end. Further, this unreal existence of ours is seen to be a condition of the intelligence identifying with it or, in other words, the present evolved condition of the intelligence of life on earth.

Since our force-line human intelligence can only process the gravitational knowledge of life at the speed of light, the residue piles up as a mighty backlog of interval or past. While life as new knowledge races on virtually unnoticed, human intelligence is almost fully occupied with trying to understand the past. And living, as the process of trying to sort out the past, takes over as the dominant activity on earth. That backlog of past has been building up since intelligence emerged out of life on earth. It has now reached such gigantic proportions that the task of processing or conceptualising it (which anyway was an impossible race against itself) has become the ceaseless and wearing toil called 'our way of life'.

211

FORCE-LINES ON EARTH

Every human being is a force line of the planet earth and his physical body is the temporary visible manifestation of the force line. However, an individual's force line is a permanent part of the earth, unaffected by the life or death of the body. While alive in the body his force line is an active circuit of intelligence converting into sense-perception the moment-to-moment knowledge of life coming down from the spirit. When he dies he retreats down his force line into the planetary psyche. Here, unencumbered by the senses, the knowledge he receives is more immediate and 'moves' much faster — ranging up to and beyond the present of Real Energy (gravitation) through to the present or presence of spirit (power). This is because behind and sustaining all people or force lines is the sun's terrestrial gravitation line represented by the body of the planet. And behind that again is the sun's spiritual power line to Draco, the serpent's head.

When we go to sleep we recede from the body's accumulated past towards the present. Sleep is merely the need to be cleansed of the past that we continuously acquire through being in a body. The further we travel up the force line towards the gravitation line, or present, the more our acquired past is dissolved and the more refreshed, restored and ready for action we feel on waking. Death is precisely the same need — to be cleansed of the entire accumulated past of that particular physical existence.

When a man withdraws into the psyche in sleep or at death, his force line does not disappear. It remains as a current of energy or life-force representing the potential difference between the two worlds of the dying living and the living dead, the positive and the negative. In this way the force line maintains the man's connection with the world, although when he is dead it has no manifested presence. Force lines are both the means of incarnation and the means by which the dead are able to revisit the earth. They also ensure that when a man sleeps or goes unconscious he wakes up in his own body and not someone else's.

Without force lines, ours would be a world in which sleep and unconsciousness were not known. A person who slept for the first time or died in such a world could never return as we do to ours when we wake up. He would find no place — no interval maintained as a force-line — for him to re-enter. The position he occupied before going to sleep would be lost; and it would be as though he had died.

In sleep the world disappears for man. But in spite of his absence from the world, time continues to release information at the same rate. This 'lost' time or information he feels unconsciously is his 'missing' past. His constant activity in the world is his attempt to find it. Each time he wakes he exerts himself in yet another effort to catch up with the past but eventually and inevitably it disappears unresolved into the mountain of all his other yesterdays. This continuous daily burst of activity to find one's own missing past is the origin of man's drive for information, his need to know what happened and why; and emotionally, of the impatience to get finished or get done. Both compulsions are as endless as they are futile, for knowledge of the past gained through concepts and the senses achieves nothing: after he rests or sleeps he has more to do and more to find out; and after that he has to sleep which puts him behind again, creating more need to find out and do.

Thus the ceaseless frenetic activity of humanity is an involuntary attempt to catch up with the information gaps created by our combined lapses into sleep, death and unconsciousness. We try to keep up by producing an ever-swifter output of information as communication. To help us we invented machines, then electrical devices, and next it will be something even more swiftly productive of information. Finally in desperation we will have to face up to the fact that it is the old story of the dog chasing its tail; that there is only one way out — to leave the world of expendable past to the robots and jump to the next time-gradient. Until we can manage to do this while alive, death does it for us now. Death is temporary respite from the hopeless human race.

THE REDUCING SPEED OF LIGHT

No vacuum is possible in the physical universe. Even light cannot travel on nothing. Every so-called vacuum — even the theoretical kind created in the scientist's mind for him to theorise about such things — contains gravitation lines. They are the medium of Real Energy along which light travels as information to form force lines.

Scientists are currently searching for what are called gravitons, the particles or waves they believe are concerned in gravitation. But gravitation is constant and, as the matrix of the universe, is everywhere and unmoving. What moves is first the data and energy 'transmitted' by gravitation and secondly the physical mind responding to the data. To date, science as a whole has shown little understanding of Real Energy and none at all of power or spirit. Its limitation is that it only recognises our time, which is simply interval or past. This omission occurs because scientific intelligence is currently grounded in the inferior planetary force lines rather than in gravitation lines. Although the intelligence of individual scientists rises occasionally to gravitational levels of inspiration, the body of scientific opinion continues to cling to the conceptualised past.

A drastic global revolution in the body of scientific knowledge is pending. It will be the biggest upheaval ever in science — and possibly in world history. A great breakthrough will follow, opening the way to a completely new system of physics for the twenty-first century. It will involve events that give an appreciation of Real Energy as I am describing it and demand a complete break with force-line thinking and intellectual materialism.

Science has yet to realise that the calculated speed of light is getting slower in relation to man's consciousness; it is no longer adequate for most advanced scientific enquiry. This is because the real speed of light is 'now'; and now is pure consciousness. As man at the top end of his intelligence returns towards pure consciousness the conceptual or calculated speed of light loses significance. As science proceeds, the discrepancy between light-speed and man's intelligence will begin to be noticed more and more. More scientific compromises and intellectual inventions such as virtual particles, black holes and Superstrings theories will have to be made to avoid facing up to the simple truth of the instantaneity of Real Energy as consciousness, with its enormous significance for man of a realisable immortality.

Man's evolution is to enable him to return to the beginning of time and the golden age of intelligence. There the speed of light was almost identical with Real Energy. Light moved so swiftly that it was almost stationary and as a result there was barely any interval or past in the world. But reasoning and rationality, the evolution of intellectual materialism on earth, put the interval in existence and slowed down light-speed (consciousness) so that now it is well separated from Real Energy or knowledge of now. In other words, intelligence separated from consciousness. The way back is for intelligence to disappear back into it.

As rationality developed, the gap or interval between now and light-speed increased. Today light-speed is a relatively slow phenomenon compared with the speed that intelligence on earth has reached at the top end of its gradient, where now is consciousness. Science will have to realise this, and face up to the implications, to enter the new-century physics.

Fourteen

THE OBJECTIVE
UNIVERSE

*We must realise that the truth of
our existence is implicit in what
we are seeing and doing now*

217

THE OBJECTIVE UNIVERSE

INDUCTION AND DEDUCTION

THE CREATION ORDER as I have described it is: will, intellect, radials of power, universal mind, space, stars, gravitation, matter and force lines. All of these 'events' occur simultaneously and together represent the moment of eternity after which nothing else ever happens except in retrospect, or by deduction.

Deduction provides our idea of time and the continuity of the sense-perceived world. However it would be impossible to describe or understand such events were it not for induction. Induction is the state of seeing the now; deduction is the method of seeing the past. Induction is probably the most poorly defined and least understood word in the philosophic glossary. This is not surprising, as induction is the act of seeing the truth or of being able to observe the moment of eternity.

Induction, being a state, cannot be learned. Deduction, being a method, is learned. The deductive process begins as the means by which we are taught as infants to evaluate our entire knowledge of experience in the sense-perceived world. We deduce everything from past information, even the existence of ourselves and the world around us. Deduction means arriving at a relative whole, or a conclusion, from a consideration of parts or details. A detective practises deduction by drawing together the clues of a crime into an all-embracing picture which hopefully points to the culprit. Even so, is the culprit ever the whole, the whole cause? Never: he is still only a part. The whole arrived at by deduction is only a relative whole, a partial whole that is no bigger than its parts. The process is selective, taking only what it thinks to be significant and rejecting everything else. Deduction is a process of reasoning in which the results can be as varied as the levels of intelligence and self-interest.

Induction, on the other hand, is a state of mind, a state of intelligence, in which the presence of the individual cannot affect the results. The results are implicit, invariable, precisely the same for everyone who enters the same state. Anyone able to observe eternity, or the now, will understand my description of both, as I will understand their descriptions. They may see other aspects as well, but those I have mentioned will be there. It is because of this that we are able to perceive, love and appreciate the truths seen by Socrates, Plato, Buddha, Jesus and great teachers of the past.

Intuition is a familiar instance of induction: in the moment I am aware of knowing a truth without having had to deduce or infer it. But as soon as I think about it, doubt it or need to check it I am back in the past — back in deduction.

To observe the now, eternity, the myth or the truth of existence, and see it as a whole, not partially, one has to be inducted into pure, objective knowledge: a state open to every man as soon as he can perceive the truth of deduction and the past. The state then implies the parts; the communication is immediate; the whole is the parts. Deduction is the other way round: from a series of parts or details one tries to infer a whole and yet can only reach a conclusion which is another sort of part. There can be no whole in the past, or coming from the past. The whole is directly experienced in the moment now — and then there is no need for deduction or parts because all is implicit in the whole being observed.

I cannot arrive at eternity, or the whole truth, by inference or deduction. I will always fall short of it — as the Big Bang theory so amply illustrates. I must begin by knowing or entering the whole now; and its parts will then be implicit as my knowledge.

To demonstrate this in a worldly and therefore partial sense: if I enter a room (the now or eternity) and describe what is there at that moment I am not deducing or inferring, not relying on past clues. What I see is implicit in my being (there). Because I am inducted, introduced into the state (the room or eternity), there is no distinction for me between the whole and its parts. Is the table in the corner not a part of the room I am describing? The weakness of this worldly parallel is that I am relying on my senses. If one of them fails me I may be in trouble, back in the dark, in ignorance. The observation of eternity does not require senses.

Induction, as the dictionary shows, is another word for initiation. When we are able to discern the now or eternity, in other words see the truth behind existence and describe or appreciate it, we are indeed in an initiated, privileged state of mind or intelligence. One is then not dependent on the past or the interval required for reasoning or deduction.

Induction is also said to be the way that laws are formulated by inference from particular experience or observations. But laws — for instance, of gravity — are not inferred; they are observed, discovered in the moment. In the moment that the law is perceived, the truth of the whole of it is present. What comes after is deduction — resurrection — and as already stated, it is always partial; for as soon as a law or principle is used deductively — that is, remembered, thought about or believed — it is no longer a whole but a part; a part of my experience, a part of my past world.

At the everyday level induction operates with the same unerring and unarguable straightforwardness. It is the act of seeing what is there in front of us now without the complication of classifying or comparing it. For example, no one can tell us we are not seeing the words on this page: the truth is implicit in the looking. All the truths of our existence now are implicit now in what we are doing and seeing. The only limitation is how clearly we are perceiving.

The error creeps in as soon as we try to interpret what we are seeing, or put it in some other context outside of the now where it is. An example is when we tell another person what we think they are seeing or should be seeing — in other words, give out our opinions. The fact is that there can be no argument with what is true now in a person's own experience: it is with his deductions (opinions and inferences) that we disagree, as that person will disagree with ours. All normal people will agree that the sky is blue if they are all looking now; but whether it was blue yesterday will be invariably disputed by someone.

We all use induction throughout our day but mostly it goes unnoticed so that we are unable to appreciate the freedom it gives. We are in the state of induction when we are enjoying ourselves or working happily or normally, oblivious of any personal problems. But when we suddenly remember a problem and start thinking, we lose our freedom and spontaneity: that is deduction, inference, establishing our identity by past reference.

Say I am in a place where I describe by telephone some of the objects around me. The game is to guess where I am. Is the game deduction or induction? It depends on which role you identify with. If you identify with the detective, the person at the other end of the phone, it is deduction. He starts off in the dark being forced to rely on reasoning, using the given clues to try to arrive at the truth of where I am now. But if you are I, the observer inside the place, giving the clues, it is a clear case of induction. I am in the room now, in the know, in the light so to speak — in the truth. I do not need clues or interval for laborious consideration. I am at the beginning and end of the game.

The question is: can the detective by deduction arrive at the truth? He might find the guilty person but he cannot find the truth. There will be other such culprits, other such relative truths.

That deduction game never ends, as life shows. What the detective is really trying to find by his life experience, and not merely by his deductive detective job, is where he himself, the man, is. Once that truth is induced, or he is inducted into it, the deduction game is up — or over.

THE ERROR OF SCIENCE

In what follows I am not attempting to denounce science, but am drawing attention to an area of conjecture into which science is unknowingly straying from the boundaries of realism. In so doing it can unintentionally uphold and perpetuate a method of reasoning in cosmic physics that is both invalid and based on false premises. As the erroneous reasoning employed looks like being repeated for a long time, it is appropriate and essential to refute it now before it infects new generations of mind. The alternative, the real, must be stated.

Man has not yet realised — and it is at the root of the whole problem of his reasoning — that he cannot calculate or deduce back beyond his own past, beyond the beginning of the species. To do so as science does in pursuing an external cosmic beginning is like solving a murder that never happened. Beyond the unique past of life on earth, there is nothing to go back to. The cosmic truth and reality are all in the now. The past is the sole, unreal creation of earth intelligence evolving through sense-perception. This is the fundamental truth of existence that has to be realised for man to jump to the next, the real, time-gradient.

Only the past, that intelligible record of the evolution of terrestrial and human intelligence, can contain differences and changes — not the now. Once intelligence understands this mighty truth — that there is no cosmic past apart from what man himself has created with his perception — the past and the value of it will be seen in an entirely new light. I repeat: all past is man's; the cosmos has none.

223

In assigning the beginning of the universe to an event in time billions of years ago, the Big Bang theory says that the universe started billions of years before now. The theory is derived from scientific observations made now being arithmetically tacked on to a non-existent 'other side' of the legitimate past, (meaning the past that has been experienced) and which began only with life on earth. The result is pure fiction.

The scientist making calculations that endeavour to exceed the human past is like the spider that spins a self-supportive web out of its own secretions. The web holds up nothing but what is essentially its own survival. In relation to the spider the web is a completely subjective structure. It is all of his own making. In both web and past there is no truth, no reality, except the man or spider shed of subjectivity — pure intelligence.

But the proposition that the cosmos has no past is naturally unacceptable. Man, as evolving intelligence, is bound to accept the unreal in all its forms until the real is explained to him and he has the chance to discriminate, or until he seriously begins the approach to reality in himself for himself.

The theoretical scientists' assumptions that the universe was at some time different to what it is now, and that change or motion is a verity, are among the several basic mistakes they make in applying the Big Bang theory. The approach would be legitimate if the theory were concerned with events within the past of the species, but it is concerned with the cosmos — which has no past and therefore is the same now as always (as the Steady State theory suggests). Now and the universe are cosmically identical: the universe is now, and now is the universe. The one self-evident quality of now is that it never changes; it is always original, always the beginning of everything. So to have any chance of being valid the Big Bang theory has to be applicable to now.

The Big Bang theorists imply that they have reached the end of the measurable universe. But every calculation they make insists on ending a split second before the beginning is reached. Science cannot get back any further, even if it appears to. The theorists calculate back to a point twenty billion years ago — twenty billion years before now — but at the same time present it as a split second away from that which demonstrably is the ceaseless origin of everything — now. Now is now. Anything away from it or short of it is clearly unreal. If just one influential scientific mind glimpsed this truth a whole new scientific epoch could begin.

But what of the split second that science cannot go beyond? Why is it there? It is there because that split second is the gap in the scientist's intelligence, the distance between consciousness and his intelligence. In that interval he and humanity deduce their universe and fail to perceive it as it is — an expression of consciousness. The split second is the condition of terrestrial intelligence itself, ceaselessly pursuing the unknown or the originating nothing — which science and all mankind is doing in one way or another. Science has unknowingly discovered the point in time of its own intelligence — a split second away from the now.

As the scientist's application of the theory shows, no calculation (that is, movement of intelligence) has succeeded or can succeed in getting beyond the split second to the nothing because any calculation (movement of intelligence) is itself the problem. The theory shows correctly that intelligence has reached a point where it is chasing its own tail. And this tail-chasing is due in the first instance to the false assumptions I am describing. It is false assumption or ignorance at every level which make intelligence move or race. Where cosmic principles are involved, any movement of intelligence is always away from the truth.

Cosmic truth is implicit in the stillness of intelligence, which then becomes pure intellect or consciousness. Cosmic truth cannot be calculated or deduced. It is always intuited, even by the scientist, and then by implication applied to formal structures. Lesser minds make deductions from these formal structures — and invariably arrive at doubtful conclusions.

When the Big Bang theory is applied historically to arrive at a supposed physical explosion creating everything out of nothing, it is hopelessly misguided and misleading. But the extraordinary thing is that essentially the Big Bang theory is an esoteric description of the true origin of the universe, provided it is applied to the universe and the observer now and not to a supposedly different universe in the past.

The theory telescopes the perceived universe into a pin-point; literally squeezes the whole physical existence of space and matter down to the size of a speck to make it all but disappear, back into the 'nothing' out of which the theory correctly intuits it must have come. It is a fantastic feat of calculative wizardry. It would be really effective, and an astonishing scientific advance, if the theorising observer realised that as intelligence he is reducing all that he is seeing and imagining, including himself, to within a split second of the non-existent now-point in his own brain, the point of reality and the emergent point of the whole universe as far as man and sense can perceive it. Then the theory would be absolutely correct. The scientist would not be concerned with a fabricated beginning billions of years ago but with himself, intelligence, being only a split second away from the pre-existent state, his own and everything's source — the motionless pure intellect or consciousness. Universe and man as intelligence would then unite in one sublime realised truth and the way would be open for scientific entry into the new epoch of time and knowledge beyond light-speed, sense and past.

The rational justification for mentally reducing the universe back to a dot is that it has changed, moved or expanded since the beginning. But that is absolutely irrelevant to cosmic objectivity; it is a subjective distortion by the observer having no meaning outside the evolution of human intelligence whose subjectivity is creating the illusion of a universe in motion. The theoretical scientist is substituting for now the condition of his own intelligence — which created the idea of an expanding universe in a purely calculated process. The truth behind the theory is that it works perfectly when applied to the scientist himself. Then there is no need mentally to change anything or to go into the past. What he is doing without knowing it and without any need for calculations is merely reducing the wide end of the telescope he is looking through to the dot of his own intelligence behind the eye-piece — which somehow, until he gets the point, annoyingly continues to persist in existing as the split second of interval between himself and reality. Man's being is that non-point, the pre-existent state of the universe. His intelligence, as we have seen, is the only moving, evolving thing, and is one iota or dot into existence from the point. That infinitely reducible, minute 'distance' or interval is represented by the split second in which intelligence surmises that it is short of the ultimate. Intelligence is the split second endeavouring to eliminate itself. It can only do that internally by reducing itself to the stillness of the intellect — being without intelligence or the need to know any more — because all is implicit in the consciousness of being.

THE TWO SIDES OF THE UNIVERSE

As it really is, the universe is stationary. As we conceive it, it is in constant motion. In what follows I explain a number of facets leading up to the actual mechanics of how evolving intelligence makes the universe move. First, let us get the stationary side of the universe clear and see how it relates to man.

The stationary, pure intellect with all its truth and power is implicit in the moment of eternity which is in every man's consciousness now. All that prevents him from inducing eternity in his own mind at this moment is his evolving intelligence; that is, his habitual reliance on his reasoning and deductive faculty. Evolving intelligence is always moving whereas eternity and the intellect that reflects it is absolutely still. The deductive faculty has no place in eternity nor any hope of ever perceiving or knowing it, let alone entering it. Fortunately, that deductive faculty or self has evolved independently of the real man that every man is. The first task now at this stage of intelligent evolution is to leave the deductive faculty in the world where it belongs and to discover the real man behind it. Man cannot induce eternity in his mind by first believing in it or deducing it exists. That approach would be back-to-front.

The real man of every man is in eternity now, is an implicit part of eternity. The real man of every man on earth is his intellect which never changes. As every man has this intellect, every man is real in so far as he is conscious of his intellect now in the same distinct way as he is aware now of his intelligent deductions (such as his thoughts and feelings) and his body. He has to discover and become familiar with the character of his intellect — stillness.

The unreal man is the constantly unsteady deductive faculty or self, the thing that thinks and reasons and that has evolved out of life on earth. The real man, the intellect, does not evolve. It is the same now as always; eternal. It is so real, close and intimate now to each man as his being that he takes it for granted, confusing it with his evolving human self whose existence is completely dependent on it. Without the presence of the intellect, intelligence and intelligent life are impossible. The reasoning deductive man could not reason or deduce; there would be nothing to bounce his perceptions off, no space in which to think, nothing to make the world real, knowable.

So we have this extraordinary situation: man's intellect is real but his intelligence is not — not until it is evolved sufficiently to give up its dependence on the past, its notions and motions, its inferences, itself. Immediately intelligence does this, it induces the now, eternity, the intellect, the real man. Immediately, intelligence and intellect are conjoined: the part that somehow appeared by reason to become separated is once more implicit in the whole, and the deductive interval — cause and effect — is dissolved or bridged.

But life or the motional world as we perceive it still goes on. How?

Everything in existence has its origin in truth. And the truth is that on the other side of existence is the intellect which is nothing. This means nothing is on the other side of existence.

The ramifications of this are enormous for anyone seeking to realise the truth.

The intellect, as I have said, is on the outer side of the girdle forming existence. Across the girdle the intellect faces itself. Nothing facing nothing reflects nothing. But as soon as existence forms and produces the evolving intelligence (man's deductive self), an observer or monitor springs up. Where before there was nothing but still, pure intelligence, there is now the movement of evolving intelligence and self. Mystics and others who realise exalted truths declare that the highest truth is nothing. In the moment of realisation they have eliminated or overcome their motional intelligent self, the monitor, and are the intellect reflecting on the intellect — nothing.

From all this arises the principle of the other.

One side of a coin (or anything) implies the other side. We deduce that the other side has the same characteristics as this side. Yet we do not know that. It is an assumption. The character of the other side of anything cannot be deduced from this side. Demonstration: the two sides of a coin, or anything, cannot be seen simultaneously. We look at one side, see what it is and assume the other side is similar. We turn the thing around and look at it — assuming we are looking at the other side; but it's still this side. In both instances it is this side we have been looking at. This side is always the sense-perceptive, universal or existential side. The other side is always nothing, or non-existent; and anyone seeing that truth, no matter how fleetingly, is looking into eternity. In that moment he is the still pure intellect. He is as nothing.

Another truth arising out of this is that the false (this side) implies the true (the other side, nothing). Our existence is false and only by seeing through it to the other can we realise the truth. In other words, what is true has to be seen through what is false.

Why can't we realise the truth direct? Because our evolving intelligence is attached to the ceaseless movement of our human mind and senses. We could not bear to perceive the intellect or man's real being as it is in its stationary state. The full blast of its naked truth and power would be too much for us. Most people are protected by the movement of their mind and senses which keeps their intelligence focused on the external world. This distracts them from any serious inner reflection which would induce stillness. But at the higher stillest end of human intelligence, the protection is the nothing.

THE SENSE-MACHINE

In the sense-perceived world the infinite truth and power of the intellect/will is broken down into objects and their movement. This is done through an amazingly ingenious and complex psychic apparatus — the sense-machine.

The sense-machine was developed evolutionarily by terrestrial intelligence working behind the scenes through the psychic brain, which I described at the beginning of the book. It is located behind the human brain and is the psychic device that makes sense-perception possible. Our senses of sight, smell, hearing, touch-feeling and taste originate there. To us they are differentiated. But deep within, in the sense-machine, they are integrated in one great sense — the sense of sense.

Our reality, the psychic double, vital being or intelligence behind our brain, is the intelligence of the sense-machine, or the intelligence that uses it. The vital being itself is beyond sense. It is immortal and never sleeps. While our sensory selves sleep, the vital being is always present, conscious. It has dimensions of intelligence beyond our imagination. Although abstractly present in our senses, here its true focus is on the reality 'behind'. It conveys the energy of this to the sense-machine.

By creating the brain and senses, the sense-machine breaks down the flow of power into a succession of pauses. Thus we receive only small and tolerable sequential doses in the form of sense-perception. The pauses however mean that our sensory world is always behind the now. What we see has already happened in reality. This means the evaluating perception in the brain is continuously trying to catch up by bridging the intervals. This makes the sensory world appear to move, giving us an attenuated moment-to-moment experience which we call time and motion.

The sense-machine and the evaluating perception in the brain, which together make up the sense-complex, operate at the speed of light. Everything relies on the speed of light for its finite or sensible existence. We must break with the habitual thinking that the speed of light refers only to the external world. Light-speed is the maximum velocity at which information or data can be transmitted, not only in physical space but also in the human mind. A fundamental error of science and reason has been to place our senses outside the laws of the universe and to ignore sense-perception as the critical factor influencing all observations and experiments. Data is data and in a forcefield such as the earth's there is no distinction between external space and mind because our minds are in that field or that field is in our minds. All data in and out of the mind travels at a maximum velocity of around 300,000 kilometres a second.

How does the vital being receive the data of reality? For this answer we have to go back to the Big Bang that accompanied the separation of the will from infinite mind. The echo of this bang is the echo of eternity. It contains all the ideas that could ever be or happen in the existence of the universe. And it is the means by which the vital being receives reality. Like a finely attuned ear, it translates through the sense-machine, what it's hearing into now. Everything in existence happens now. Now is the moment of eternity in the terrestrial mind and in sense.

For us, however, reality is the earth-idea. The earth-idea contains all the ideas that make up our earthly existence and is held in the seventh level of terrestrial mind — the intellect. In the girdle of eternity that forms existence, intellect and will are in perfect union and form one great being. This being may be called the Lord of Existence. Being all and containing all it is completely self-sufficient, the supreme truth. There is no movement or time in the great being.

As all the ideas of universal existence are held in the intellect of the great being, so all the ideas of earthly existence are held in the terrestrial intellect. Also, through the sense-machine the eternal echo of the one great being is translated into myriad sensations of little beings — us. Similarly, the truth available to the vital being is also available to us. Thus, a man can become so sensationally stilled by being detached from the senses that he may look back and away from the sense-machine direct into the nothing and through the nothing to reality. There, from behind the nothing, the intellect reveals its truths through realisation, intimation and myth. Yet, no matter how precise and amazing is a particular realisation of truth, the highest truth is nothing — and is known to be nothing.

THE SOLID IN WHICH WE LIVE

Reality contains no parts, no separation and no interval. It is total and complete togetherness. So how is it that the physical world, the sensory version of reality, consists wholly of separate bodies and objects?

The 'transmission' of reality to the vital being is instantaneous, no interval is involved; then the sense-machine translates the reality at light-speed into the physical world. There's a huge discrepancy in time between the two functions. This accounts for the sensory perception of space, the overlooked phenomenon that links us all. If we look closely at the physical world in front of us now we will see that in the whole of what we are apprehending no separation exists — provided we see space as part of the whole.

When the reality of the earth-idea is converted into sensory images, the sense-machine pulses the images back to reflect off the terrestrial intellect deep within. (Every sense image of what you are seeing now in the world has to be affirmed by a matching idea in the intellect.) The image is then reflected back to the brain enabling it to comprehend the perception. However, by the time the image covers the distance out to the intellect and back at light-speed, the moment of the brain's apprehension of the sensory reality is a micro-interval behind the moment of sensing.

This discrepancy between the real and its sense reflection is tiny but intolerable. It has to be compensated for under the universal law of conservation of energy. The compensation is made by a shrinkage of the sense-image. It returns to the brain smaller than it was. Thus, every object we perceive in the world is 'smaller' that it really is. The combined shrinkage of all the images put together is the space we see. Space is sense devoid of image. And that space is what joins everything together just as surely as if all the objects in the world were their original size.

This also partially explains why objects in the physical world appear larger when closer and smaller when distant, depending on where the observer is. But there is another factor: the degree of reality of the object being observed and the degree of reality of the observer himself. Evolution, cosmic and terrestrial, is the process of things (ideas) and observers (intelligence) becoming more real. In cosmic evolution — I am now talking of evolution outside life on earth — the only real objects are, first, the stars, and then their planets or other orbiting matter in which life is potential or already exists.

The correlation between size, distance and reality can most easily be demonstrated by looking at the sky and stars. The greater the reality of an object as seen from the earth, the smaller it appears and the more space is created between it and the observer. This is because the gravitation of a star exerts such an enormous pull on the sense-machine (which is hard up against it, not distant) that it reduces the energy of the pulse carrying the image back to the intellect. This makes an even more drastic reduction of the image received by the brain and a correspond-ingly acute creation of space. As image reduces, space increases.

Near a star, space is intensified and very different in quality to the space near a less real object such as the earth or an earth observer — 'less real' because stars are more real than orbital matter and its intelligence. Also, near a star there is a sharp intensification of time which because of our distance we cannot appreciate. Thus, although stars appearing to be at the greatest distance in deepest space have the smallest image or effect on the brain, the effect on intelligence, the evolving state behind the brain, is inversely tremendous.

Intensified space, or stellar gravitation, is the universal power that draws life in the form of intelligence out of surrounding orbital matter. (This is what happened to create life on earth.) As the life-form develops sufficient intelligence, intensified space around the stars attracts first the attention, then the interest and finally the wonder of intelligence (in our case man). Then the intelligence begins to reach out, to speed up or attempt by investigation to travel back intellectually through the interval of the past and space towards reality. Simultaneously, the truth is released by intensified time — the reduced interval or past that occurs in the vicinity of stars for someone approaching reality. This provides the inspiration that allows intelligence to discriminate between the motionless reality it is trying to approach and the force of the distracting, incoming light-speed stream of information from the world. Against this the intelligence must persevere to remain still, detached and undisturbed.

In other words, intensified space draws out intelligence; and intensified time provides the cosmic knowledge (Real Energy) it needs to resist being weighed down by the past and engulfed by the incessant onslaught of sense-information.

THE LIVING PAST

We have seen that the delay involved in the sense-image travelling from the sense-machine to the intellect and back to the brain means that when it returns to the brain as a perception it is always old. Due to the finite speed of light a micro-interval occurs between the original and the perceived image. As this micro interval shrinks the image and creates space, so it creates past. Each piece of space carries an indelible impression of the original image 'erased' from it by shrinkage. These permanent spatial impressions through endless, countless perceptions since 'time' began, or since the first perception, is what the past is. These accumulated, indelible impressions form a record of every moment of the past, stretching back to the very first life-forms on earth — the great memory of the human race. This racial memory is our own living past.

Man can only begin to understand this living past, or have access to it consciously, to the degree that he succeeds while alive in uniting with his vital double in the world of the living dead. This is done by endeavouring to clear his own personal psychic space as I have described earlier in the book. Although the racial past or memory with its enormous range is beyond the recall of man as he is, he reflects off the substance of it all the time, to affirm his external sense impressions. For instance, if he sees a tree the external image is reflected off the image of 'tree' in the past and he then 'knows' he is seeing a tree. This is what is called 'being objective' and 'sticking to facts'.

Without this interval of past we could not reflect as individuals. Selfconsciousness would be impossible — we could not recognise our own existence. In fact, without the past there could be no life as we know it.

So, due to the accumulation of this racial interval or memory, our brain now not only receives and perceives the physical sense world but is able to reflect back the other way on to the permanent racial record of it — something like another world — and by making comparisons between the two, evaluate the present, reason about the future and deduce the past.

Within the same racial memory the individual man gradually builds up his own personalised impression of events thus producing his personal memory. While he remains objective or factual he keeps this at the level of the racial memory and it works fairly well and efficiently in his personal life. But when he reflects exclusively on his personal memory the result is emotional reflection containing hardly any intelligence. The necessary false, through which the truth can and must be seen, degenerates into blind ignorance which can see neither the false nor the true.

All sense information is past and represents the false through which the truth has to be seen. But the false is not ignorance. Even terrestrial intelligence or reflection which provides us with the physical sense-world has to be termed false because it consists of past images of reality. But terrestrial reflection is valid and real compared with the deluding information we receive from reflecting exclusively on our personal memory.

Intelligent reflection on personal memory consists almost entirely of ignorance moving at the speed of light. The consequent error and distortion are so enormous for man's self-conscious intelligence that they create most of the tension and confusion in the world. The result for the individual is spasmodic worry and severe emotional instability which the unselfconscious species do not have to contend with.

BEING OBJECTIVE

What is it to be objective — really objective? How can we perceive the world in an entirely new way, for that obviously is what it means.

Normally, to be objective means being able to perceive a thing 'as it is' without the observer allowing emotional considerations — his subjectivity — to influence the result. That is sufficient to arrive at and deal with the facts and practicalities of life. But to arrive at the truth of both life and death — to realise and participate in a greater reality than is normal — requires a more fundamental objectivity that amounts to a revolutionary change in the individual's way of perceiving.

When we are being normally objective how can we be sure

there is no subjective factor already distorting what we are seeing? We cannot. In fact and truth, everything we see — the whole physical world — is already subjectively conditioned when we apprehend it. For one thing, as I have explained, no object in the world is its original 'size': all is hugely diminished by conversion to sense, which simultaneously creates space (another distortion) and also past (a further distortion). Moreover, due to its present evolutionary status, our terrestrial intelligence, as it makes sense of the sense-world, creates the false impression of movement — another distortion. Our normal perception is like a four dimensional movie that embraces and encloses us, and we cannot get out of it except in a very special way — by realising objectivity.

What humanity regards as being objective, even at the professional and scientific level, is still a phase in our innate subjectivity. However, behind our normal awareness, is a real objective point which we have to discover, reach and realise. Objectivity is so subtle and elusive to our subjective comprehension that it defies definition: we cannot be told in advance what we are looking for. Hence the need for it to be individually realised, in the same way as being in love or any other sensational state has to be realised before it can be truly known.

The first step towards realising objectivity is to understand subjectivity — the almost total subjectivity of human perception. This we will now attempt to do.

BEING SUBJECTIVE

Three subjectivities are at work in the human psyche. They are: first, the senses; second, the egoic naming process; and third, the emotional self.

First, the subjectivity of the senses.

The senses present precisely the same physical world to all humanity. Here there is no thought or calculation, no re-arranging of context or significance outside of what is; no 'this' or 'that', no 'me' or 'mine'. Here is the impersonal human condition where all humanity (the observer) is fundamentally one in the one world or perception. Mountain or tree appear the same to all. At this point of human awareness, no persons exist. There is no evaluated difference in what is seen or who is seeing it; no adopted position or attitude. The observer could be anyone, is anyone. Here, in fact, we are the impersonal intelligence of humanity observing the world as it is from a myriad of individual positions in the human psyche — hence, terrestrial intelligence.

However, as straightforward as this initial sense-perception seems, it is still, as I have explained, subjective.

The second subjectivity is the egoic naming process.

This almost instantaneous process is where humanity starts to divide into persons. Here our individual or divided self begins. The world, having been perceived as the senses present it to all humanity, is then taken over by our individual perception which starts the attribution, orientating and naming process. This is the mental dynamic of existence. Although not a part of the senses, or sense-machine, the ego attaches itself to the senses and by the naming of objects and repetition starts constructing a subtle, intellectual memory. By ceaselessly re-identifying things through language or thought it creates a conceivable, considerable, reflective world. This intellectual world is the basis for our reasonably presumed professional and scientific objectivity.

Naming requires past. For a thing to be named or recognised it must have been previously observed. That initial cognition and essential past are provided by the senses which act first by conveying and reducing reality down to our perceived physical world; this shrinkage of the sense images in the human psyche provides the necessary space and past in which the ego can set up shop and start work.

The work of the ego, which begins even before birth in the womb, is to fill the child's psychological space with its own images or personal impressions of the perceived world, thus gradually building up the subtle memory. In this task the ego is

238

assisted by the parents' (and environment's) continual repetition of naming and reaffirmation-by-experience of identifying words. Thus language and thought gradually become substitutes for objects — and the intellectual world is created.

The original reality converted by the senses to the world of what is, is now further converted into the world of what was — the memory. Here the ego lives and operates ceaselessly as thinking, calculating and dreaming. The ego, being outwardly attentive on the world, has — in spite of popular misconceptions — a minimum of selfish identification. It is only very faintly dependent on the emotional body (the third subjectivity). The ego is like the flame at the tip of a wick that will continue to burn with very little wax (body or emotion).

The third subjectivity is that of the emotional self.

The virulence, potency and violence behind all human subjectivity is in wanting which comes from the emotional self and expresses itself through the physical body. As the ego is the self of awareness and thought, this emotional self is the self of feelings. It is the real weight behind selfishness. By pressing up on the ego, it uses it — forces it to express its desires in words and thought in the same way as the wax of a candle 'forces' the wick to keep burning. The emotional self creates the personally exclusive and highly viscous memory of 'me and mine'. This memory is not intellectual but rooted in feelings.

The emotional self, as previously explained, is formed out of the psychic reservoir of past and self created by all life that has ever lived on earth. This vast reservoir of past — the racial memory — is pathological, and I use the word in the sense of the original Greek root 'pathos' — suffering. It includes the original, experiential memory of all the species which is preserved in the deep unconscious of the terrestrial psyche and is expressed in the intermediate levels as man's emotions. The racial memory is incredibly residual, immemorially ancient. Through each person's emotional self the whole mortal past of the species, their sufferings, dreads and fears, influences the behaviour of humanity in the now. Backed by this immense

239

force of sensuous past, predating thought and enlightened knowledge, the emotional self compels the ego, when its desires are opposed or crossed, to name, accuse, judge, and argue with vehemence. Words and thoughts are used as weapons or lures, depending on the kind of sensational or physical gratification this emotional subjectivity demands at the time.

Those are the three subjectivities separating us from reality or objectivity. And yet, paradoxically, they are the only means by which we can attain to it. In the following section (as in this book generally) I will be speaking to the first and second subjectivities in the reader because the third, the emotional self, unless quiescent, has no chance of or interest in understanding or participating. Only in the ego's realisation of objectivity does the emotional self start to become enlightened. Even then it is a very long process to bring the power of realised objectivity down into the emotional self and into the body — which is the point of life after death and life after realisation.

MENTAL ASSOCIATIONS

As the ego, we are unable to make sense of anything in isolation. Each of us perceives individual objects not as themselves (as the senses present them) but in association with other objects; that is, partially and historically. Everything is placed in a context of the past, grouped in the subtle memory with other objects previously named and remembered which themselves had been subject to the same time-slotting process right back to our earliest infantile perceptions. Everything is strung out along the line of time, in the past, so as to be understood. That means giving each object or condition a selective relationship to other objects which the thing itself does not have. By doing this we create reasonable significance — one thing pointing to another, cause and effect, continuity, or in other words, historicity.

Whether we are awake or asleep the ego is ceaselessly active in our memory, slotting and re-slotting even the most trivial perceptions into some 'reasonable' context. With the habit of years it gains an insane sort of tempo like a chattering idiot whose aimless jabbering is often heard on the edge of sleep. Outwardly it is repeated in the aged and senile when the restraints and need of practical performance have broken down. For all of us, what began as the necessary business of establishing a functional memory degenerates sooner or later into an involuntary nightmare — the uncontrollable restless mind.

To begin perceiving objectively we have to release ourselves from the instinctive associative and grouping habit. We have to start observing objects without subtle connecting thought to the past or future, and without being afraid — for if we succeed, even for an instant, we are on the edge of eternity where nothing familiar in time exists and awful loneliness or emptiness can be sensed. It will be obvious from this that thought, as the associative function, is time.

To start being really objective we have to learn to interrupt the automatic naming or thinking process. We must thrust our intelligence in between what the senses present and the normal aimless, egoic reaffirming of what is seen. The ego must no longer be allowed to identify things habitually and think indiscriminately in the subtle memory as it does all the time when we have nothing particular on our mind.

This means that for a long while ahead we can never relax as we have in the past. We must be uncompromisingly vigilant, not allow one aimless movement of thought in our heads. We have to stay in the now: see only what the senses are presenting every moment without judgment or subtle discrimination, and denying any memory reference whatever unless it is to find a fact to carry out a practical action. We must commit ourselves to not discussing or relating past events.

We will fail frequently. But we must not be discouraged, for even to be discouraged is a sign of egoic interference, reference to the past. Life is now and for the purpose of discovering its

241

objectivity, its deathlessness, we will never run out of time. We must be patient and resolute, never naming failure but forgetting it by staying in the now and trying to do better next time — which is now, anyway.

We have to be very honest with ourselves. We seldom need to refer to the past, even to yesterday. If we do, we are probably fooling ourselves — or the ego is fooling us. In the first twenty-one years of our life we have done most of the naming we ever need to do. What we have to do now is dissolve the thick layer of past that has built up over and obscured the impersonal point in our psyche through which we must go forward into increasing objectivity, or reality, finally to arrive at the one real undivided and undividing being that each of us really is.

PERCEIVING OBJECTIVELY

To be objective in any instant we must be free of thought or evaluation. We must resist the pressure to run the instants together, to think about or consider what our senses are reporting. We must see with our senses without naming what we are seeing, without imparting historicity to the object. We will thus deny the object subjective continuity and remove it from the past and future, as well as from ourselves. We will then be perceiving it exactly as it is.

But to try being objective with any object or person in our daily lives is extremely difficult. Each person or object has too much habitual association for us. As a result of our constant, reaffirming, associative thinking, the familiar world around us has solidified into blocks of time from which it can be agonising to try to pry ourselves free. Even to lose one piece of it, such as an old friend, someone we love or a valued possession or position, can produce great distress and grief. It is the memory of things lost that pains us so much. This memory reflex extends down into the cells of our bodies. Addiction is a cellular craving for repetition or continuity of the familiar, the known, the past.

To start to perceive the objective reality of existence we have to exclude those things that are sub-effects; that is those things that are not cosmic, that were not present in the first instant of eternity. This of course includes all man-made objects, as well as people as both represent successional time or interval and to begin with are almost impossible to perceive objectively. If we go outside on a clear night and observe the stars without any reference to the earth, or speculative thought, all that we see is what manifested in the first instant that never ends. As the universe we are looking at is now, it is as it has always been. It is the beginning and the end. It is there when we are born; there when we die. It is eternal.

If our perception is very swift, we will realise that this instant of observation is identical to that first eternal instant, and that each instant in which we fail to see any motion in the heavens is the same timeless, eternally present instant. By looking at something that has no past, such as the cosmos, and seeing it as it is, the observer himself sheds his past, loses his subjectivity. If, on the other hand, he thinks, calculates or draws a conclusion, he has awarded it a past it does not have; he has imposed his functional self on it, given the timeless a piece of his own limited intelligence, and therefore has rendered the objective subjective.

Inasmuch as a man thinks and calculates, he is seeing a function of himself; and the reality remains as distant and elusive as ever. The universe is actually a mirror of himself until he learns to see through it and its appearances. Later, when he can look at any object or person without subtly imputing past to either, the same universal reality will be perceived.

In observing the heavens correctly we will see that the universe is always as it is, irrespective of sub-effects such as people, objects or circumstances which may appear in the scene. Objectivity is to hold this timelessness of the now, irrespective of what else may be going on 'below' in human affairs. If in our daily lives we can remain identified with a timeless object like the universe, or the now, we ourselves are similarly timeless and therefore similarly objective, eternal and real. A man is as mortal or immortal as the idea or time he identifies with.

If when looking at the heavens we are tempted to surmise that the starry scene we are observing has changed since the first instant, we are speculating and thinking successionally. I repeat, now is the first and only instant. Whether or not the objects we are seeing have changed appearance or position is irrelevant and dependent on thought and time as interval. Such reasoning requires two instants to be run together — as now and then — and therefore we enter the time-trap of subjectivity. Let us say we are observing the heavens and we become aware of movement such as the twinkle of a star. This twinkle is a sub-effect caused by the earth's atmosphere (really, our own subjectivity) and as such is not part of the wider universe. We have to look through the twinkle, through the effect, the sense-appearance — and see the star.

The starry heavens we are observing is the field of objective reality. What it consists of — matter, energy or whatever — is irrelevant and to attempt to name it even scientifically is to fall into subjectivity. That field of reality cannot and does not change. The condition of the field, its sense-appearance at any time, is secondary, a sub-effect, created by the oscillation of evolving terrestrial intelligence. That intelligence, it will be remembered, is behind the initial point of sense-perception — before we individuals come into existence — and is the evolving intelligence of humanity as a whole. It is the relative unsteadiness of this intelligence — its relatively unevolved state compared with the finest intelligence in the universe — that creates the moving sense-perceived world. The uncomfortable but unarguable fact is that any change or motion we perceive in the cosmos occurs in the past, in ourselves. And of course the past cannot be the present — let alone the eternal instant, or objective reality.

To higher intelligence, motion as we see it hardly exists; and where we apprehend relative motionlessness such as in the perceived heavens, higher intelligence is aware of fantastic activity. And with that perception our intervals of centuries — which are measures of past, not time — pass in seconds.

When humanity evolves into higher intelligence the sense-perceived world will not necessarily disappear. We will just elect to spend less time in it, as in dying we involuntarily spend less time in it now. We will then not merely be able to see through movement and sense-appearance but will begin to 'pass through' both so that they are no longer barriers to a greater, more universal existence.

Fifteen

THE FUNCTION OF
THE INTELLECT

*Looking objectively at evolution on
earth we can now see that it is the
process of making the terrestrial
intellect fully functional*

THE FUNCTION OF
THE INTELLECT

EVERYTHING IN EXISTENCE comes out of the eternal intellect. And the eternal intellect replicates itself down through all the worlds to the physical. Although degraded by the descent through the psyche and into matter, the cohesion and harmony of the reality is relatively maintained and holds each of the worlds below together. Our immediate demonstration of this is the cohesion and harmony of nature.

Below the eternal intellect is the stellar intellect, the intellect of the Lord of Existence. Next from our point of view is the solar intellect. This is followed by the terrestrial and finally the human intellect.

The human intellect consists of all the materialistic concepts we have formed of the world; and all our emotional and false impressions of life which men and women have held to and repeated from generation to generation for thousands of years.

It could be called a subconscious common memory but in relation to truth it is a garbage tip. In material terms it includes our concepts (degradations of ideas) to do with money and business, theoretical science, recreational drugs and the practical activities that built the world as we know it. In emotional terms, the human intellect includes the notion (also the degradation of an idea) that the death of the body is the end of life, that love is personal and a personal feeling, that beliefs are real, that there is virtue in worldly position and power over others and so on. The continual use of the human intellect gives rise to insecurity, worry, fear, violence and self-doubt, particularly where love is concerned. The reason for this is that the human intellect is the creation of man and the intellects above are part of the cosmic order of creation.

Everyone using the human intellect will at times reflect off the terrestrial intellect. These will be moments of unfamiliar clarity and intimations of a greater reality. But usually they will be quickly covered over by emotional feelings, thought or talk and the virtue of the moment lost. By constantly reflecting on the human intellect man reaffirms its erroneous notions and materialistic concepts and so perpetuates his personal stress, worry, fear, conflict and anxieties. For here is repeated once again the cosmic law that says nothing can exist unless it is perceived or acknowledged by intelligence. When man thinks negatively or emotionally his intelligence is affirming those negative notions in the human intellect. Similarly, when man concerns himself with purely materialistic concepts and drives he affirms his vulnerability to loss and the misery of dying. With most men doing the same thing most of the time, the notions have become fixed and substantive — and a deeply destructive way of living.

The spiritual or holy life begins when man starts refusing to give in to the almost irresistible urge to identify with the human intellect. By not allowing his intelligence to reflect on its negativity, the destructive emotions and thoughts start to disappear for him. He engages in the same practical activities perhaps but without seeing them as an end in themselves. As this happens he begins to see through to the terrestrial intellect and reflect off that. And from there it's only a matter of time (perhaps a long, long time) before he is reflecting on the solar intellect as well.

Although the human intellect is practically all negative in relation to truth, it has its positive values such as kindness, generosity, compassion, patience and giving. As worthy as these qualities are in our sensory existence, they are nevertheless degradations of the idea of love held in the terrestrial intellect. Real love, living love, is all that is good. There are no other words for it; it embraces and replaces every seemingly virtuous description.

The ideas in the terrestrial intellect relate to truth, beauty, harmony, order and rightness. By regularly perceiving and acknowledging these, a man activates them; and his life starts to reflect that harmony. But since most men seldom have time for such acknowledgment, the ideas remain weak, not easily discerned and therefore fail to change the life of the majority. When an idea is not used sufficiently it starts to fade and lose luminosity (in the same way as a negative emotion). The more it is used or acknowledged the more luminous it becomes for the individual and for all men using the terrestrial intellect.

The terrestrial intellect is not personal as the human intellect is; it determines our lives behind the scenes. The scenes are merely the emotional, ignorant and largely unhappy world created on the surface of existence by the personal human intellect. Impersonal means 'having no personal consideration'; all is determined according to the eternal laws of justice and truth. Being impersonal, the terrestrial intellect is the keeper and sustainer of each individual's force line, which, as we have seen, controls the moment of his birth and his death.

THE WORLD OF FIGMENTS

No man has a personal intellect, so no man is a person. A person is a figment in the imagination of the human intellect. The human mind, being a figment, loves figments; whereas love —

love that does not waver or vary — loves the individual being behind the person. We remember the figment, the person; not the individual behind it.

The individual behind the person is a part of the terrestrial intellect, that part of the terrestrial mind containing the evolving soul or psychic double at Level Four. Not everyone living has an evolving soul. Many are figments. Their reality is so limited that they virtually disappear from the after-death evolutionary cycle; they remain as the force-line that merely maintains their position. They are more a position than a being.

Throughout the day while alive and awake each individual reaffirms his position in the world. At the same time he affirms the particular part of the human intellect containing the figment of himself. He does the same for others, by recognising and considering them, as they do for him. Thus everyone's figment remains relatively discernible in the human intellect and imagination. So the person can be imagined when he is not there. The same applies to someone who has died. If they are frequently remembered and thought about by many who knew them, their figment will be kept more or less intact in the human intellect. But if they are truly loved in their absence, that is, loved without the necessity of thinking about them, the impulse of love will go as a benediction to the impersonal being in Level Four — because love is the greatest power behind the evolution of the soul.

Famous dead people known only from books and the reports of others are figments in the imagination of perhaps millions of people who have different figmental images of them. Although the figments are totally without reality, they remain in the human intellect where all falsehood is gathered for new generations to think about and add to. The original figments are eventually almost unrecognisable — as has happened to those of the Buddha and Jesus, due to the enormous imaginative activity of their followers. However, the reality of a famous person, if he has evolved sufficiently, is in the terrestrial intellect, where no thought or ceremony can reach him — only true love without thought or imagination.

Humanity's ceaseless daily perception of the objects in existence — things like buildings, roads, people, nature and even the moon — is what prevents the world around us from starting to disappear. This is the cosmic law again: unless intelligence reflects on an idea or an object, the idea or object begins to disappear. While one individual continues to use the intellect in this way, the world as he knows it holds together. But if he alone existed while the rest of humanity went to sleep for a thousand years, they would wake up to a very pale and scarcely recognisable replica of the world they used to know. Through lack of use the world-ideas would have faded in the intellect and almost have returned to a state of pure potentiality. As it is, with less than one-third of humanity asleep at any one time, we all wake up from sleep (and, one day we may discover, from death) to a world that has continued in our absence in a solid, sensible way.

THE ACTIVATION OF IDEAS

The evolution of life on earth is the process of making the terrestrial intellect fully functional. For the most part the intellect is potential. Every idea in it is there; but to come into existence or appearance each idea or part of an idea has to be activated through being perceived by human intelligence. In this lies the difficulty. The necessary effect of human intelligence is to downgrade an idea; ideas are reduced to concepts, notions and words and they lose their original virtue. There is then a feverish movement to fill in the idea in greater detail with concepts, theories and descriptions. For any idea to retain its virtue, it must not be broken down and degraded in this way. It must be lived as my life. But the living of ideas such as truth, love, beauty, harmony, divine order and rightness is very, very rare.

For the majority of people, living is the process of filling in ideas in more and more detail through the human intellect. This goes on with ever increasing gusto. An example is the intense concentration on numerous ideas in the fields of biology, medicine and technology. We know more detail today than ever before about practically everything, from insects and plants to the human anatomy and space rockets. And we are extremely busy filling in every idea that catches our personal interest or fancy, from astrology to meditation methods to innumerable therapies. Here the terrestrial intellect above overlaps part of the human intellect below. All the ideas in the terrestrial intellect are there for the spiritual and physical health of man. Inasmuch as man applies the ideas to that end they have virtue. But if they are pursued for intellectual satisfaction or selfish reasons they are without virtue; and the man concerned misses the opportunity to deepen his enlightenment. Service to humanity without personal consideration is love. And love is the finest virtue.

Often in man's endeavours there is a mixture of virtue and other considerations as he activates an idea for humanity to follow and improve on. The conquest of Mount Everest in 1953 is an example of the initial activation of an idea accompanied by the virtue of courage and persistence. When the will embraced the idea, man reached the top. And from then on other men were able to fill in the idea with increasing ease. Since the first ascent, several hundred climbers have reached the top — and one even without the use of oxygen! The idea of the conquest of Everest has been so filled in that even tourist parties will soon be managing it. Again, the four-minute mile was established for the first time in recorded history by a runner in the 1960s. It was hailed worldwide as an incredible feat. Today scores of runners complete the mile in well under the four minutes. As the idea continues to be filled in they will go on doing better and better.

If an idea is superseded and neglected it starts to fade towards dissolution. This activation and deactivation of ideas — what we call today's fashion and custom and yesterday's out-of-date interests and activities — is the dynamic behind our intelligent evolution.

The familiar objects and conditions around us now were not always familiar. Each had to be cognised and activated in the intellect by one man's original perception so that other men could re-cognise and use it. In this way humanity's perception is continuously broadened.

At first, an original idea may be completely incomprehensible to most of humanity and remain dormant for centuries. But having been activated, the certainty is that sooner or later it will recur in man's perception and eventually enlighten him, despite the ignorant resistance of the majority. There are numerous examples of this in history. One is the Copernican idea that the earth revolves around the sun, an intuition that had occurred as an activation of the intellect 1800 years earlier to the Greek philosopher Aristarchus. Two thousand years later, and two centuries after Copernicus, this great truth was still not luminous enough (or conversely, human intelligence was not brilliant enough) for it to be received by humanity; Galileo was forced to recant and deny it in order to save his life. An idea activated in the intellect can be said to become more luminous with use (or, intelligence, more brilliant) and more accessible to more of humanity until it becomes an 'obvious' truth to all. Unfortunately, this usually means that the idea is superficially known but seldom understood with the brilliance that such great ideas as the Copernican insight should evoke in man.

But for the individual man there is much more to this activation and acknowledgment. It concerns the beauty and fulfilment of life for him. Although the flowers and trees, the sky and the rain and the whole of nature is substantively present due to the activation of these ideas by generations of the whole of humanity, they remain largely taken for granted; a backdrop to his worldly activities. He goes for a holiday and hopes the weather, the accommodation, the scenery and the circumstances generally will be favourable. They may or may not be; it's a chance he takes with most things most of his life. His life is a sort of gamble that everything will go okay once he's made the formal arrangements. But life in truth — the idea of life — is not like that. Life lived in truth is the continuous acknowledgment of the beauty and rightness of life in its innumerable natural forms. Acknowledging with gratitude the flowers, the birds, the grass, and particularly the people who are loved, ensures that the individual is accompanied by all that is good and right. Finally, there is no need to take holidays; a much better equivalent than could be planned is provided. And if at the same time thanks and gratitude can be given to the integrity of life behind the scenes — not as a supplication but as a selfless, spontaneous offering — that is the most potent power of all, for the good of all.

Will man ever succeed in filling in all the ideas in the terrestrial intellect? No. That would be the end of his competitive existence; there would be no more to know and nothing for humanity to do but enjoy what was already there and provided. That is a definition of heaven; and the great bulk of humanity, at its present level of intelligence, doesn't want anything to do with it. Nevertheless, the more enlightened very small minority has reached that level of intelligence. These people can be said to have activated enough of the terrestrial intellect to be capable of reflecting on a section of the solar intellect. The intelligence then has access to more abstract knowledge leading up to the realisation of the transcendent spirit or God as love and truth; and in a more formal sense is potentially capable of comprehending the presence of extra terrestrial intelligence, and of communicating with it.

This all means that man, as intelligence on earth, now has potential access to that vast and profound area of the solar/stellar intellect represented by the Draconic Transverse, the section spatially hundreds of light years across containing and controlling the life and death cycle of human existence. It is extremely doubtful whether any form of intelligence in the universe as we know it is capable of using the entire universal intellect. However, some cosmic intelligences are so mighty that they are able to use the vast tracts of the intellect represented by whole galaxies. Eventually, as man gains access to wider or extra-human sections of the intellect, using his more mature reflection on life, death and the other realities behind the perceived universe, he will one day activate sufficient of the terrestrial and solar/stellar intellects; and this will coincide with some extraordinary cosmic discoveries and events.

Sixteen

THE MYTHIC UNIVERSE

*Before going further into the
reality behind existence, here
is an overview of the mythic
dimension of the universe*

THE MYTHIC UNIVERSE

BEHIND THE HUMAN MIND

THE MYTHIC UNIVERSE is behind the universe we see with our senses. It is also behind our concepts of the universe and the innumerable scientific theories about it. In other words, the mythic universe is behind the human mind and is the source of all existence.

Only myth can describe what is behind the human mind. Science has no myth so science cannot do it. Religious myth has endeavoured to. So have the myths of the ancients and the primitives. Each has told its story of what's behind the universe. And the stories are as numerous as the stars in the sky. That's because, in the mythic universe behind what we see, every star represents an idea in the mind of the Great Being — God.

But this Great Being, which I described earlier as Lord of Existence, is only the ultimate God, not the absolute God. As we have seen, infinite mind out of which will and intellect emerged to form the idea of universal existence, is closer to being the absolute. But is it the absolute? Or is infinite mind still an ultimate to the unspeakable absolute God whose absoluteness is the guarantee of never-ending life?

Such talk is mythic. The scientific mind cannot stomach it. But to the man behind the scientific mind and the man behind the human mind, such mythic talk can have a strange uplifting quality — uplifting of course towards the most high, the absolute God. Where else is there to go in a life that never ends? Only in the scientific mind and human mind does life end and therefore have no enduring purpose.

Myth has no before or after, no past, no future. We can never establish to the satisfaction of our rational minds that myth is true or real; we can only believe it or not believe it. The contemporary demonstration of myth in action is the appearance of UFOs. UFOs do not come from the outer universe as is popularly believed; they come from within — from the mythic universe. At a speed far in excess of the calculated speed of light, they emerge from the depths of the universal mind, enter the Draconic Transverse, then the solar and terrestrial minds and finally sensory existence. Here, due to the resistance of the past, their speed is vastly reduced; as is their ability to stay visible for long. Being mythic, meaning really non-existent, they have to consume the immediate past around them to exist. As soon as this is exhausted they vanish.

The mythic universe consisting of eternity and every possible idea of existence is contained in the universal intellect. This is a totally abstract and potential world beyond all other worlds but containing all other worlds. If all other worlds in existence vanished they would continue energetically in the mythic universe of the intellect.

HOW IDEAS BECOME MANIFEST

When an idea comes into existence it appears in the sensory heavens as a star. Mythically, however, that star has a mind and intellect of its own. Our star is the sun. So there is the solar mind and solar intellect. All the ideas in the solar mind, of which the earth is only one, are stored in the solar intellect. All intelligence in the orbiting planets of the solar system must first use the intellect of the planet, then the solar intellect.

A cosmic idea such as the solar system sits in the universal intellect for eternity or until it is ready for existence, which means that the idea has been embraced by the will. Even so, the idea of the solar system could not appear in sense because the law is that nothing can manifest in our sensory world unless it is observed by intelligence. That happened the day that primitive man saw the sun for the first time, as described back at the beginning of our story [see page 22, 'The dawn of self-consciousness']. At the same moment he cognised or 'knew' the earth in the form of his immediate environment.

The rest of the solar system, and the earth, gradually came into man's cognitive and sensory existence; as the other planets did thousands of years later when observed by his intelligence. Before that, like now, the solar system was present in the mind in its entirety; but only those parts have manifested that his intelligence can observe. Man deduces that he discovers these things but in truth and fact he manifests them.

Today, a great deal of the solar system and what it contains has not yet been manifested; it waits for man to see it. This means that even now the solar system only partly exists. It is all there, but it is not all there — not until man cognises all its parts. This is the mystery and the drama of his evolutionary journey which will take him deep into the physical universe. It is also the cross-over point between the mythic universe and the physical, the point where nothing becomes something and something disappears into nothing. It is the point that science has reached and where it hovers frustratedly on the edge of matter or sense. But while science remains materialistic or orientated to matter it cannot cross into the mythic.

THE TIMELESS

Man cannot truly begin to understand the wider universe and participate in a greater reality until he has grasped the timelessness of the solar system, or the difference between the cosmic and the cosmetic. The temptation is to imagine that there was a physical evolutionary beginning to the solar system. But such a beginning implies past and the solar system, being intrinsic to the universe, has no past. The only past in the solar system is that which developed with life on earth. Anything that is not now as it was in the beginning, consists of past. Certainly we can see evidence of innumerable changes on the earth since the start of the species. That in fact is the trail the past has created and left behind by the passage of our evolving intelligence. But planet earth has not changed cosmically; with the other planets it continues to circle the sun with the same unfaltering precision. The sun pours out its heat into the universal matrix of gravitation; and all that is cosmic is majestically untouched. The changes we know or see are purely cosmetic, transient. The changes change again and keep changing. Nothing within our purview is stable or predictable — except the timeless rhythm of the solar system that has no past and therefore knows no change.

The solar system's physical existence is part of the phenomenon of terrestrial life. It did not precede life. It was not 'here' before life started. Life 'brought' the solar system into existence. As life became selfconsciously intelligent it gradually filled in the idea of the solar system until we see it as it is today — partly formed. It was not the solar system that evolved but life's intelligent ability to make form clearer and more intelligible. Take away life and both the physical solar system and universe disappear.

The sun and planets we see, like the rest of the visible universe, are projections by the sense-machine of the reality deep within the unconscious. There the real universe, as the eternal or universal intellect, is as it is, and not as we sense it, conceive it or think to change it.

THE LIMITS OF INTELLIGENCE

The universe, and therefore the solar system, are completely responsive to the evolving, ever-expanding demands of intelligence on earth for more knowledge. What intelligence needs to know next is always there in the moment — but not necessarily what it wants to know. For man today, the physical universe is still not as it really can be; it is only as clear, or complete, as his intelligence can conceive of it and that perception will be different again for future generations. The perceiver, intelligence, changes; not the objectivity perceived.

Intelligence is the one factor which prevents the immediate realisation of the unity of life here and everywhere. Human intelligence, for instance, is currently bound by sensory perception which produces different bodies that conceal the unity of the life within them.

The intelligence of sensory perception is always busily active, as we can see reflected in the ceaseless movement of things. This activity is actually the evolving of human intelligence. Only by movement can human intelligence eventually learn the value of no movement. As this occurs, the intelligence begins to separate from the busy ignorance of the human mind and becomes more aligned with the terrestrial intellect which (to us) is absolutely still. With increasing stillness the intelligence is enabled to perceive beyond sense-perception to the greater unity behind it. Then, as thought, calculation, speculation and interpretation cease and the intelligence becomes absolutely stilled, it begins to reflect off the pure, stationary, solar intellect. Cosmic life with its other intelligences is realised. Distance and the physical universe as separating factors vanish: all is life, this indescribable being that I am.

So, between life elsewhere in the universe and life on earth lies the impediment of our intelligence. The whole physical universe, inasmuch as it attracts our speculation and scientific investigation is in fact a massed condition of materialistic ignorance hiding the reality of the mythic universe behind it.

While man is entranced with delving into physical causes he misses
the mystery of life. Ultimately, all life is the one life. And ultimately,
all that life anywhere can succeed in achieving is to observe, know
and finally realise itself — that is, its own inseparable oneness —
even across the vastness of the physical universe.

THE COSMIC BODY

Intelligence is cosmic. And being cosmic it has no past. Since
essentially we are intelligence — universal intelligence — and
not physical bodies, we are not individuals. We are individual
intelligence only so long as we think we are. In our part of the
universe, where life exists as humanity and all the species living and
dead, we are the combined intelligence of our cosmic body, planet
earth. Each of us is a cell in the body of that great cosmic being.

Our evolution is fundamentally the cosmic evolution of the
earth. The earth itself is perfect but the intelligence that has
emerged from it — us — is not. In cosmic evolution, the earth
has its unique place and 'task' in the hierarchy of cosmic bodies;
its intelligent life must as closely as possible reflect the original
perfection of the idea of the planet. The earth must contribute
the particular divine quality the earth represents to the illustrious
being of the cosmos. That quality is love. There is no love in the
rest of the universe.

Each cosmic body has its special quality to contribute —
qualities beyond anything we can imagine. From this we can see
that wherever intelligent life develops in the universe it is a
reflection more or less of the divine perfection of the cosmic
body or original divine idea from which the intelligence
emerged; and not of any apparent individual part. And yet in
cosmic consciousness that part is as great as the whole — is the
whole, when the whole is realised.

Looking out at the vastness of the physical universe from the
earth's position would be an extremely chilling if not devastating

experience for any individual who for a moment connected with the terrestrial mind and realised the planet's cosmic isolation in space. This awful, inspiring aloneness is sometimes realised by man. In that moment he exceeds the limited individuality of the part and shares in the beauty, precision and rightness of the whole.

THE LIMITS OF SENSE-PERCEPTION

Although we are able to observe the formal appearance of other stars, and as cosmic observers contribute to their continued existence, we are forced to settle for a subjective, mainly calculated or imagined scheme of the reality in which we live. The reality of our solar system is of course mythic, not sense-perceptive. But that hasn't dawned on our general intelligence. Our entire knowledge and existence come from sense-perception; and as creatures of the past we can only think about or know what is already known. Everything 'out there' is of our perception — just sensory images created by our sensory brain. The stars we see are not what the stars are or how higher intelligences see or know them to be.

To try to look at the universe from the point of view of other intelligences of course we have to apply our sense-perception; and as no other sense-perceptive life exists in the universe it must give us a totally wrong impression. Sense-perception is dependent on the speed of light whereas higher intelligence operates at the infinitely superior speed of now — no time. This means that humanity cannot hope to see as higher intelligences see — unless we cross into the mythic perception and get the idea behind the words. Nevertheless, applying sense-perception to higher intelligences, we can say that the physical existence of the solar system could only be seen by them in its cosmic entirety from astronomical distances away. The greater the distance from us (from our intelligence) the greater the reality.

The reality of the solar system is not perceivable to life within the solar system or even to life immediately outside it. To us and those immediate observers (if they exist) the solar system is not yet complete, not yet real; the idea is still mainly in our imagination and only exists in fragments. We can observe most of its planetary parts only in isolation, never as a whole, and we depend on our imagination to put together what we think it is. Our imagination of the solar system is intelligence working in the past to describe a cosmic system that has no past. So our imaginative images and concepts are nothing like the solar system really is. We are cosmic babes.

Cosmic or spiritual consciousness is liberation from the burden of the past. This does not mean that the sensory world ceases; it merely enables one to see through it to the other, the mythic reality behind.

We cannot see our place in that reality until our intelligence can 'travel' outside it to other star systems by transcending the confining past of the human psyche. We can do this only by learning to see life more objectively, without self-reference or past reference; that is, by learning to see through the impediment of our evolving sensory intelligence which is so concrete, materialistic, and past-bound. When we manage this we will be cosmically conscious, able to participate in the knowledge of the cosmic reality from which we come.

THE MOVEMENT OF INTELLIGENCE

The scientific Big Bang theory of an expanding universe requires a centre for the bang to have occurred in. But there is no such centre to the universe. The only centre is the position of the observing intelligence. Thus, wherever intelligence (the observer) is, that is the centre of the universe.

From that point, the universe as knowledge is always seen to be expanding. But what expands is not the universe but man's knowledge of the stationary intellect. Remember that as we observe the starry universe 'outside' we are simultaneously reflecting off the universal intellect 'inside'. The intellect absorbs and always will absorb the attention of intelligent life wherever it arises in matter — until the intelligence (the observer) realises the truth of life within itself. The universe is not expanding, as science suggests; it is exactly the same as it 'was' in the first instant. Nothing is moving except intelligence.

In relation to the instant in which each section of the universe appears to the observer, there was and is no first or last, oldest or newest. Such concepts of time are gained from the sensory point of view. The universe of galaxies is the sensory view of what in the mind is the stationary universal intellect.

The nearest point of reality for the observer is the star of the system to which he belongs — or, in the case of 'travelling' higher intelligence, the star of the system in which he happens to be. Any apparent movement of stellar objects is due solely to the instability of the observing intelligence. The stars stand for all the ideas of reality and, as human enquiry demonstrates, eventually intelligence always comes to reflect on the stars. Each star is an idea, a concentration of information. Each piece of matter in the universe represents the potential emergence of life and intelligence in time.

All intelligent life in the universe shares the universal intellect. Supra-intelligent beings existing elsewhere in the universe are reflecting off the same intellect as us, only more of it. In our case we are limited to the terrestrial and solar parts of it until such time as we exceed them.

As intelligence observes the stars (and obversely, reflects on them off the universal intellect) the ideas in the stars devolve into information. Stars devolve, knowledge expands, while intelligence increases. The information is transmitted back off the intellect at light-speed to become man's perceptive awareness. The calculated speed of light, being a finite yet absolute velocity for the transmission and receipt of sense-perceptive information, is the factor that creates the impression of motion, or interval, giving rise to historicity or our concept of time and past.

Meanwhile, supra-intelligences use star power to journey or swing from system to system on the mythic side of the universe. This is a way of being — swifter than the speed of light. Ultimately you can only travel in your own being.

Seventeen

THE GIRDLE
OF ETERNITY

*This chapter outlines the system
of being that we call reality*

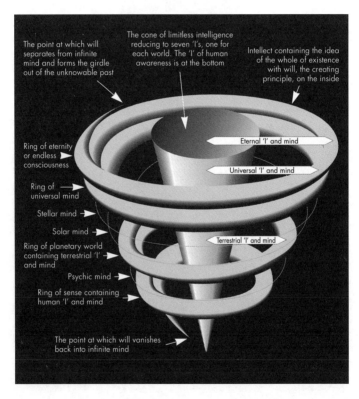

Spiral girdle of eternity in infinite mind

THE GIRDLE OF ETERNITY

THE SYSTEM OF BEING

THIS IS THE spiral girdle of eternity. It is all there is. It sits in the unimaginable nothingness of infinite mind, unfathomably deep in the mind reading this.

The girdle was formed when the will separated from infinite mind, bringing with it the idea of eternal existence. In an instant — the moment of eternity — the will formed the spiral consisting of every idea that could ever be. It then vanished back into the infinitude whence it came. The illustration shows the spiral's beginning from nothing, the infinitude around it, and its ending in nothing.

The girdle consists of the will, the intellect, consciousness and the seven worlds and minds of existence, the last and least of which is the human mind producing the physical world at the bottom of the spiral.

The cone-shaped axis is a zone of limitless consciousness which descends through seven reducing levels of intelligence or I, one for each world.

271

The top first loop of the spiral is the outer gallery or ring of eternity embracing endless consciousness.

Within and below the first loop is the second gallery, which encircles universal mind.

Then comes the stellar gallery containing stellar mind; and after that the solar gallery containing solar mind. These are 'spatial' and not 'solid', so they are indicated in the illustration by broken lines.

Next is the solid gallery or ring of the planetary world containing terrestrial mind.

Within and below the ring of the planetary world is the psychic gallery, or world of the living dead, (another 'spatial' gallery indicated by broken lines).

And finally there is a solid gallery of sense, containing the human mind and the physical world.

The entire spiral girdle is a massive, self-sufficient, objective system of being and existence in the infinitude of mind. It is powered or fuelled internally at every level by the consciousness and will of eternity — the now. Its cosmic, spiritual name is reality.

REALITY IS AN OBJECT

With our imprecise knowledge of the human psyche and even less perception of the minds above, it is not easy to conceive of reality as a precise and defined structure. Being spellbound by the world of the senses, and by a seemingly external reality, it is difficult for us to realise that everything including the universe is actually in the mind and that all is contained in one single idea/object called reality — or the spiral girdle of eternity as depicted in the illustration.

Here on earth, man is on the bottom spiral of the girdle and from here he has two ways of looking for reality. One is to turn his attention inward and look back through the human psyche, up through the girdle towards the solar and higher minds above —

that is, through the object of reality itself. The other is to look 'out' at the apparent reality of the universe around him. He has been looking out through his senses in this way since time began; but as nothing in reality can exist outside the end of the spiral where he is, the external world that he perceives is only apparent, a relative reality which affirms its transience or unreality by disappearing every night when he sleeps or when he goes unconscious. Consequently, in looking out through the senses he has never succeeded, nor can he ever hope to succeed, in perceiving reality.

THE MIRROR OF INFINITE MIND

If the physical universe is not real, how does it appear at all, and appear with such convincing reality? The answer, in the first instance, lies in the extraordinary quality of infinite mind surrounding the spiral. Infinite mind cannot be 'seen through'; it defies perception. It is totally, absolutely, reflective. It turns the observer's vision, the eye or I, back on itself so that the observer sees in the reflection where he is now in the girdle of reality, which is where he is in the physical world. In other words, as man's place is at the bottom end of reality, when he looks 'out' at the infinitude of mind, he sees a reflection of the reality behind him — which then appears to be in front of him but in physical form! It is like holding up a mirror and looking into it back over your shoulder. The scene in the mirror appears to be happening in front of you when really it is only a reflection of what is happening behind you. Further, what is approaching from behind seems in the mirror to be actually coming towards you from the front.

You are seeing the physical world around you in that way, at this moment. But the world around you is a world of sense. And in reality there is no sense to be reflected, and no sense in infinite mind. So the world of sense must be an effect of

273

something between reality and infinite mind. It is. It is an effect of the sense-machine, which, you will remember, is in the psyche behind the human brain. The sense-machine is a psychic apparatus that makes our physical world — makes sense — although in itself it is an abstract part of reality, which means there is no sense in it. Every moment the sense-machine receives the ceaseless flow of reality from 'behind'. This conveys the abstract data of the idea of existence to the sense-machine which then converts it into sense and projects it 'out' towards infinite mind.

But something else precedes this. Remember, you are looking into the mirror of infinite mind. And before any sense is projected, the mirror reflects the abstract reality streaming down from behind you. In the mirror the reality is coming towards you; but it is imperceptible as yet because it does not make sense. It is purely a reflected stream of power. The sense-machine projects its sensory interpretation of reality into this incoming stream of power. The meeting of the two provides the dynamic for the sensory appearance of existence. This enables the physical brain with its mind and five senses to 'make sense' of reality; for without translation into sense, reality would have no meaning.

If you are perceptive and still enough, you can sense at any moment the reflection of the current of reality streaming down from behind and keeping the whole world together and moving in the mirror. The stream is now — the now that never stops or ceases for you, the same now that fuels reality at every level.

As I said, man, deluded by his senses, burrows into the mirror image looking for reality and has never found a trace of it. Reality is the other way. By turning his attention inward and going with the incoming current of reality, he finally meets and realises the stream of reality coming down the girdle. He travels back towards the unknown, against the flow, like a fish swimming upstream. Given sufficient will, he passes through the sense-machine and beholds reality direct behind the sensory delusion.

The man travelling inwardly retains his sense-perceptive existence because although the sense-machine cannot project inwardly it continues to operate outwardly. It needs resistance to produce the positive reflection of the physical universe, but a man going with the incoming stream fails to provide the same resistance as previously. As a result he begins to have two simultaneous perceptions; one of the projected sense-world around him (to which he is losing his attachment) and the other of increasing consciousness of the reality within. His attention continues to refine until it is indistinguishable from the down-coming stream and finally, at death or before, he disappears into the solar consciousness above.

Reality itself is not a creation, not an effect like the physical universe. The universe, in spite of its grandeur, is only the sense-machine's partial, projected approximation of the downcoming inner reality which never changes and is absolutely self-sufficient. Take away the senses and the real universe within is unaffected.

While alive and awake we earthlings must always remain sense-perceptive beings — mirror-beings. However, in time we can and must introduce the perception of reality into our sense-existence through the inward-going journey of the attention. We must remember that reality is always within, that it cannot be registered outside through the senses. It can be reflected through the senses, through the universe, which accounts for the beauty and wonder we perceive, but it cannot be held as an enduring awareness until the attention is sufficiently refined to focus on, or join, the point of reality itself — where the mirror meets the real within the individual.

LOOKING TOWARDS INFINITY

In terms of the seven levels of terrestrial mind, man must look 'back' from Level One, where he is on the earth's surface, through the psyche (Levels Two and Three) and the higher levels of mind (Four, Five and Six) towards Level Seven. Although the end of the terrestrial mind, Level Seven is also the beginning of the solar mind; and the end of the solar mind is in turn the beginning of the universal mind.

A man looking inwards through the minds in their correct order is actually looking 'up' through the girdle's central axis of intelligence. Then the minds and each I function like a series of lenses, each of greater resolution, which finally enable reality to be seen as it is. On the other hand, when man looks out from the earth's surface and perceives the external universe through the senses he gets precisely the opposite effect, like looking through the wrong end of a telescope. The cosmic bodies seem very distant, removed, remote, when in reality they are not. Looking out like that or through astronomy or space travel into the incoming stream of consciousness, man can of course never hope to get back to the beginning, reality itself.

Probing into the incoming stream is the beginning of infinity. Infinity — represented by the formal universe — arises when one perceives against or travels out against the stream. In truth, as well as by definition, infinity can have no end in itself: it is endless. Having no end, infinity like the universe can have no beginning outside of the man searching for it. Such a man looking for the reality of himself in the infinite universe finds only the endlessness of the world. All of this means that as a physical being on the face of the earth, and at the tail-end of the incoming solar/universal stream, man actually occupies the last outpost in the infinite energetic universe. In the future, wherever he travels to in outer space, that place will still be yet another reflected part of this reflected 'end' and therefore continue to be part of the endless journey into infinity.

Man just cannot escape from infinity in the external way that he is pursuing. His assumption that he should be able to take the obvious physical route straight out from the earth into the reflected solar mind, or solar system, is understandable. But the error in this reasoning is that man is not really a physical being. The fact that the individual physical man's life on earth is very brief, and that when he dies he is never seen again, is pretty convincing evidence of this. As well as his physical being, man is, as we have seen, both a sensational and a mental being able to reflect the higher levels of terrestrial mind — though he is not yet a solar or cosmic being. Otherwise he would not be at the tail-end of the solar reflection, confronted with nothing but its infinite forms and concepts. He would have a knowledge of reality. He would be at the beginning within himself, where the one idea of reality is complete without concepts.

The prospect of immortality

Man's physical occupation of the tail-end of the universal mirror-image (the beginning of infinity) is as if he had invaded an alien cosmic territory scourged by a disease called Death. For the human race to retain a hold on the earth, man's individual physical selves must keep on dying and being born to comply with infinity's demand for endless cycles. But man can make this otherwise alien territory his own. By discovering the reality of himself within his physical being, he can bring immortality into this physical life, and death the scourge will then be conquered, overcome. Immortality is man's own real state anyway, and by bringing it consciously with him into his mirror-life on earth he completes the mysterious, sacred circle, linking the beginning (reality) and the end (infinity), uniting Yang and Yin, as fore-shadowed by that most ancient of spiritual and occult symbols — the ouroboros, the serpent swallowing its own tail.

An amazing life lies ahead of human beings when physical life on earth is no longer interrupted by death or made superficial and self-serving by the dread of it. Man will then be responsible for this part of the cosmos and his life will be filled with extraordinary new purpose. But for now he has to be told that immortality exists, and intellectually persuaded of it, so that he can start breaking down the barriers of attachment and the preoccupation with his mirror-self and mirror-world.

When a man dies he does not go out into solar space, into the infinity which he strives so hard to penetrate while alive. He withdraws back into the psyche into his own real world of Level Three, behind the outward-going drive of matter and sense. There, right beside him, is Level Four, the beginning of the higher terrestrial mind and the resplendent centre of the world of the living dead. Depending on how much his psychic double has been purified or made luminous by its projected life on earth, and the man has been purified by going inward with the stream, he can then look straight through the higher levels of the spiral girdle to the solar/universal minds and see reality as it is. It was with such splendid inner visions of reality that the great prophets conceived of paradise, heaven, or nirvana.

THE GALLERIES OF REALITY

The spiral girdle of eternity as I have described and illustrated it is the home of all intelligence and life in the cosmos. This ranges from the highest principle of intelligence above the universal mind — God — down through the great spirits of the stellar and solar mind to the terrestrial mind and to man, who for the most part is floundering in the inner and outer, life and death dilemma of the psyche at the bottom of the girdle.

The girdle is nevertheless our reality, no matter how lowly or tentative our position in it may seem. The individual man's intelligence can very swiftly rise up the central cone. As he starts to ascend out of the misty human subconscious into the clarity of the terrestrial mind, he begins to realise that reality is indeed his, that time itself is not against him but at his disposal, and that reality is a wonderland, a substantive objective idea through which he can wander at will because it is his home.

Each of the minds in reality has a gallery. Each gallery is a unique world as well as an observation port from which the individual intelligence can look out at the surrounding infinitude of mind as from a space station. Later we will look around each of these galleries and observe the girdle from each vantage point, but for now I will briefly indicate the quality of the two galleries to which man has nearest access.

The gallery of our physical world is the projection of the sense-machine. Its observation port provides the familiar view from the surface of the earth. When we look out we see the reflection of nature — rich and beautiful but mortal and ephemeral for the dying living. When we look or go in behind the sense-machine we discover the energetic reality of nature which is shared with the living dead. Together, these two hemispheres form a whole, representing the incredible future fullness of immortal life which it is humanity's evolutionary task to bring about.

When a man's psychic double is clearly and correctly aligned with the higher minds he gains access to the terrestrial gallery of the spiral. This gallery connects with the inner radiant world of the planet, the life splendid — immortal life already realised. Looking out from here into infinite mind, or contemplating the void of nothing, one may energetically perceive the entire spiral girdle of eternity reflected as in a mirror. When observed through the terrestrial gallery as a reflection in infinite mind, the girdle of eternity has an astonishing quality. Not only is it possible to perceive reality as it is now — it is also possible to see how it began, and all its constituent parts still in the act of forming.

Eighteen

SOMETHING
FROM NOTHING

*The story I am now going to tell is
literally a journey into the
consciousness behind the surface
mind — an introduction
to reality*

SOMETHING FROM NOTHING

THE RING OF TRUTH

HOW CAN YOU tell if the story which follows is true, or real, other than by perceiving reality for yourself? The only verification you can have that it is true and that there is such an amazing object as reality in the mind, is if while I am describing it, you are able to discern in yourself a sense of correctness, rightness or recognition — in other words, the ring of truth. The ring of truth is none other than the ring or echo of eternity, the sound of reality itself.

In this journey and story each man is truly his own authority. To follow, you do not need to understand all that is being said or have to try to remember any of it; only to listen with an open mind. The truth in you will do the rest. It is your own mind-structure I will be describing and so you should recognise at least some of what is in it. And some parts will seem familiar to you because in telling this story I go over some of the ground already covered in the book, but now from a much deeper and even more profound point of perception. From the very first page we have been going deeper and deeper into truth or reality, and steadily approaching a greater consciousness.

The fruit of this journey is self-knowledge. By voluntarily going with the stream into the structure of reality through an open mind, you release powerful new energies in your unconscious. To the degree that a man is enlightened by such energies he is prepared for the state of conscious immortality at the inevitable moment of his physical death, when he disappears back into the psychic reality whence he came not so long ago.

THE ECHO OF NOW

The will separated from infinite mind, bringing with it the eternal intellect containing the complete idea of existence. In that instant reality began to form. The separation of will from mind caused the mightiest and most enduring bang of all time — an echo that continues in our mind today as the now that never ends, and the sensation we never cease to feel. That echo is a complete reproduction in sound, percussion or vibration of everything the intellect contained; therefore everything that ever could be. In some ancient religious philosophies it is described as the 'Word' or Logos and in another as Aum, the sacred sound of creation. All existence following the echo would be merely a reverberation in a lower octave of this one eternal instant which by us must be regarded as absolute timelessness — eternity.

The continuous moment of the now is to us the optimum point of the present. But being an echo it remains a reproduction and therefore a past expression. Hence, as everything that follows occurs within now, all existence happens in the past.

The now as it emerged in that original first instant was similar in character, but very different in nature, from the now of today. Now, as then, retains the same absoluteness in that it cannot be lost, delayed, outdistanced or avoided. It sweeps on inexorably in life, death and unconsciousness. It is the one demonstrable constant of all time. But in that first instant before any worlds existed, or any I was present to perceive them, the now in its absoluteness contained all space, time and matter. It was all there was, an inconceivable block of unknowable past — unknowable because no I was there to observe and know it.

THE EXPANSION OF SPACE

Today, the space we think and perceive in is the remains of that original solidity of absolute past. But it has been rendered insubstantial by the extraction from it of all existence in all the worlds to date, under the observation of the I of consciousness in each of the worlds. For the cosmic rule still applies: nothing can exist until observed by intelligence, and as no I had yet arisen there was nothing but the inconceivable block of original past.

The process that creates space between objects in the physical world (described on page 232) is a replication in the dense sensory octave of how space is created in the non-sensory higher worlds. Space is essential to every level of the girdle of reality. But higher space, even in the next world above ours, is utterly different to our space. There it is timeless; there is no interval, so there is no distance and no separate objects. Every energetic part is linked by the space to form the one consciousness. Although above that, space is incomprehensible, every level continues to be a refinement of the one below it. Higher space is indescribable; but a description in sensory terms can enable the reader to grasp the idea behind the words.

So reality 'runs' on the fuel of the now; existence is the positive result and space the negative remainder. As more and more existence (knowledge) is extracted every moment from the now, space increases proportionately. As we discover new matter such as stars or extend our knowledge to include considerations of the moon, planets or the subatomic microcosm, we become aware of more space, time and distance in the universe. Similarly, as we add to existence in the form of time-saving technological advances, we again create additional time or space for further activities of existence. We cannot help making more space and time because space and time are the matrix of our existence. Moreover, every day the solidity of the individual's own existence is minutely eroded and replaced by more space. This he feels as a certain emptiness in his life or himself, often inducing him to question the purpose or fear the uncertainty of his life.

The fact that man is more and more busily engaged in trying to fill the world's expanding space, or his own inner emptiness or loneliness, is incidental. The truth remains that space (universal and personal) is growing as it has done ever since existence first began in that all-substantive moment of eternity. But for man it is now compounding at such a rapid rate that there is less and less fabric to existence: what was yesterday the established norm of social or civilised interaction is today considered to be disintegrating and fragmenting. Deep personal uncertainty, which is impossible to allay by communicating or sharing old emotions, as in the past, affects all but the very young and the very insensitive. The need for real solutions, original foundations which only the truth can provide, is considered to be crucial. Materialism, the intellectualising of matter or the world as practised in theoretical science for example, eventually will no longer be enough for people. Intellectual materialism is too tenuous and abstract to support those frequent moments when the pressures of existence and personal emptiness are upon us. Rational thought is merely mentalised matter and it is useless as a support against the expansion of space and the increasing personal diffusion which that causes.

THE ORIGINAL WORD

The inner emptiness so often felt by people these days is reflected in the emptiness of our language. Most of the important words we use have lost their reality — as we have. The original power of the word to communicate true meaning is mostly missing. We seldom use words rightly or objectively. A right or objective word is a word which retains its original cosmic, spiritual or pastless energy; therefore its correct signification. When correctly received, right or objective words do not allow the play of subjectivity. Such words include: eternity, void, I, power, now, love and God. Today's language has invested these words with meanings totally out of character with their original significance. As a result they have lost their power to inform us fundamentally.

Most words in all tongues today relate to human experience that occurred long after the events which created the impulse of language. In our passage through historical time away from the truth of the beginning of man and his language, right words have become encased in layers of subjective and relatively meaningless meanings. Meaning is meaningful only inasmuch as it relates to origins. From the beginning of language, human subjectivity has gradually attached itself to words and spawned generation after generation of derivatives until few if any words communicate original meaning. This blanketing subjectivity derives from imprecise usage arising out of assumptions, ignorance and aimless chatter; together with considerations of the market-place, self and personality. Language as now used is mainly a record of man's subjectivity and no longer a demonstration of original truth. Any such truth in it is well and truly buried.

Originally, human language was not just a tool of convenience or conversation. It arose out of the need to preserve in existence and to communicate to all generations the original truth or original events which gave rise to existence — such as the contents of this book. These events were all present in the mind, in its pre-existent nucleus — the moment of eternity. A word originally did not symbolise an event. It contained the event. The effect of a word in the recipient's mind was to produce an insight into the moment of eternity where the event the word signified was (and is) continuously being 're-enacted'. Put another way these events are forever 'recurring' in eternity — in the mind of us all at this moment, now. All we as individuals have to do is to get there in our own consciousness by absorbing these words, and perceive for ourselves what is happening there.

Once, every word was an aspect of eternity. Instead of merely having a meaning, each word provided a miniature demonstration in the man's own consciousness of its eternal significance, followed by what today is called 'understanding' (knowledge without memorable knowing). This almost extinct process of systemic comprehension existed before subjective memory developed in man. Today, subjective memory is the principal obstruction to this amazing original way of apperceiving and demonstrating the truth.

Psychedelic drugs such as LSD carry a similar energy to objective words. When taken into the physical system they produce equivalent effects, which might be called living archetypes.

Since original, objective words relate to the beginning of existence, the start of the world, all of the world's earliest or root languages necessarily begin with creation myths. The difficulty today in relating to the myths (and incidentally in relating to drug-induced images) is that our interpretative faculty is subjective, whereas the original words or ideas of the myths were, and still are, objectively present in our consciousness. Stripped of our acquired and contrived subjectivity we would immediately perceive and comprehend their meaning. The all-powerful deterrent to this com-prehension, is that when correctly interpreted the creation myths (and the drug-images) demolish the rational subjective world we have erected with our conceptual language. Man can rarely perceive the truth of the myths — let alone live this truth — without seeming to lose touch with his reason and his world.

Original language was sparse. Words that comprise most of today's language were unthinkable, utterly unnecessary. To the intelligence of man at that time they would in fact have been unworthy of being included in language because the import of such words was already obvious in the action and process of life itself. Tree, for instance, was not a word for the reason that it was a living, objective fact and had externalised for all men to perceive for themselves. Words were only for the signification of ideas that had not yet manifested as objects — eternity, void, I, power, now, love and God. Objective language demonstrated the inner life, as nature and activity demonstrated life that had externalised. Thus for a long, long time there was no need for other words. All existence was self-demonstrating, inwardly or outwardly. There was no artificial world which language has since created. Original ideas had not become conceptually adulterated because tomorrow, the first word signifying escapism through artifice or man's flight from death, had not been thought of. There was no death so no fear and no need of any tomorrow or of an unreal world built on words or concepts. To the intelligence of man at that time language as it is today with its host of inferior words would be regarded as completely redundant, serving only to demonstrate that the men using it had neither correctly perceived nor understood life as it is. Such men — ourselves — would be regarded as completely subjective creatures living inside a word-spun hypnotic dream. The language of all races today is almost entirely superfluous to the truth of life. This is a mighty thing to realise.

Original number

During the very early evolution of language, an alternative was developed — mathematics, the language of science. Men whom we might call the first scientists, or who were the first to demonstrate a scientific inclination, responded to the increasing disappearance of right words and the proliferation of conceptual or subjective language and tried to trace their way back to the historically receding truth, the original event or moment of existence. This they attempted to do through using external symbols, instead of through themselves — as was the inclination of the men demonstrating divine or true philosophic consciousness.

Casting around for an objective system, these original scientists seized on 'number', which was then emerging from language as an abstraction for measuring the environment. Impartial, non-subjective number, they reckoned, would substitute for human language and allow them to get back to the beginning of life. They were doomed to failure from the start. Even though number was evolving out of language at this very early stage in human history, it still came later than the word. There was no chance of number revealing the beginning; it would always fall short. Number is only relatively objective and therefore liable to subjective distortions of time and position. Also, the rot of subjectivity had already set in and become part of man himself. The original truth, simplicity and purity of man and his original language had already been substantially overwhelmed.

Mathematics can only ever lead back to the point of existence where numbering began; near the beginning of phenomenal existence, but not to the beginning of life. Thus, as the calculations of the Big Bang theorists confirm, science cannot use numbers to get back to the moment before existence externalised out of nothing. The zero and one minus one, or something minus something, are still symbols representing the concept of 'no thing' — not the reality of nothing. The symbols are in space and time and are still within subjective conceptual play. Right words are not symbols like mathematical signs. They are the real thing and they go back to the state before the beginning of existence, the moment of eternity.

THE PRIMARY PROPOSITION OF EXISTENCE

The idea of the world or universe coming into existence necessitates nothing to manifest as something. In the approach to reality this can be called the primary proposition of existence. Implicitly, it demands a demonstration. The quest for that demonstration is the perennial search for truth and our origins by science and true philosophy.

As we have seen, science cannot clear the jump back over and into the nothing. It falls down this side of existence, a split second before the primordial Big Bang. Theoretical physics, the esoteric wing of science, nevertheless continues the attempt. Its way of demonstrating 'nothing becoming something' has been to invent something where nothing apparently existed and call it a 'virtual particle'. Although admitting that the virtual particle has no existence, the physicists have found its presence essential for the articulation of their theories of the creation of matter. Yet the virtual particle does now exist. It exists because the physicist has made it exist by discovering or perceiving it in the mind. At the edge of reality and existence where nothing is continuously becoming something and something continuously disappearing into nothing, all is occurring in the mind of the observer able to penetrate to this region. The more one-pointed his attention, the deeper and more confidently he can penetrate into the clear abstract areas of mind; and the more he perceives originally, profoundly and necessarily mythically. Thus even science has had to rely on a semblance of myth, the last resort for all who would seriously attempt to articulate the truth behind existence.

Although serving to demonstrate very simply the point at which something becomes nothing, the theoretical physicist's method still does not reveal the enabling principle behind it. Like a man demonstrating running by running, he merely demonstrates invention by invention and discovery by discovery — all aspects of existence — and takes for granted the enabling moment in which the demonstration occurs. The fact is that nothing can become something, or something nothing, only because of the now or moment in which these and all events occur. In the approach to reality any demonstration, even the demonstration of existence as living, is secondary — unless it reveals the truth of the now.

The now is the point of reality behind the something. It is also the point of reality behind the 'no thing'; that is, behind the space that remains when something disappears. In other words, space, as well as the objects and demonstrations appearing in it, are all phenomenal; they are on the material side of nothing. The task is to get behind space itself and into the nothing; and the only means is through the now. Consequently, the next crucial step in the approach to reality by both science and ourselves is to penetrate the now, to go into or look into the heart of the myth: into the moment of eternity. The moment of eternity — now — lies between existence and reality. There, and only there, in this never-ending moment, can what is behind 'no thing' be perceived.

THE TRUTH OF THE BEGINNING

The moment before the beginning of existence is deep within our own mind. For our mind is simply an extension into sense of original infinite mind. By keeping the mind still and focused on what is being said, we can energetically participate in the truth of the beginning. This is because the beginning is not in the past; it is happening forever now within us.

We are looking into the void of infinite mind, into nothing. We are immersed in it. We do not know or need to know whether we exist or do not exist. This state is like being conscious in the stillness and silence of deep dreamless sleep. Suddenly a tremendous crack of 'sound' fixes our attention on a point somewhere in the void. This is the echo of the instant of eternity, the now, approaching and expanding towards us at absolute velocity.

Absolute velocity means absolute mass which is equivalent to absolute substantiality and solidity — primary energy at its unimaginable densest. The sense-perceptive effect of the now emerging from the void like this would be as if the faintest distant star suddenly expanded at such incredible speed that in a single instant it filled the entire field of perception and engulfed the observer. But this does not happen sense-perceptively, although it does happen in eternity. In eternity the solid point of the echo, or now, expands in a flash, filling the whole void, including our watching consciousness, with that incredibly dense mass of energy — original time or past — and it leaves no interval for any further existence. Although our consciousness remains completely unaffected by this enveloping energy of solid time, it can be said that the energy may be apprehended as an unspeakable blackness, darkness or oblivion. Nothing exists or could exist in such primal density. Apart from the presence of the observing consciousness, all opportunity for further existence is past forever in that single instant.

However the paradox of eternity is that what happens there also does not happen there. In eternity everything is cancelled out and nothing — the truth — remains. Yet for existence to exist there has of course to be 'something'; and we are endeavouring to observe how the something arises.

The now cannot continue as infinitude in the presence of any thing. So as the point of the now emerges at absolute velocity, the watching consciousness instantly retreats or withdraws at infinite speed. The effect is similar to that of a spaceship suddenly accelerating away from an approaching, apparently expanding planet and by its superior speed 'reducing' the planet's size and maintaining it at the size of a dot. By exceeding the absolute speed of the now, the consciousness escapes, transcends the absoluteness of time, and vanishes.

This induces the first time-change in eternity, out of which arise the three principles behind subsequent existence: intelligence, space and energy/matter.

First, intelligence. Out of the vanishing point into which the infinitude of mind as infinite consciousness retreats and disappears, I arise. I am the first condition of all existence — intelligence. While consciousness alone is non-existent (yet ever-present in the oblivion of original time and of what does not happen) I who arise out of consciousness am the first effect of time outdistanced or transcended; hence I am the beginning of existence. Stated another way, consciousness is the presence in the non-existent stillness of what does not happen; and I am the presence in existence of all that does happen.

293

Primal space (negative vacuum) is the second principle of existence arising out of the eternal time-change. It determines the subsequent character, the 'no thing-ness', of universal space; or the positive vacuum in which existence occurs. Primal space, or the negative vacuum of eternity, is created by the absolute expansion of the now (that which does not happen but happens in eternity) being made abstract by the withdrawal of consciousness. Into this primal vacuum, which may be described as the interval of eternity, the point of the now emerges as 'solid line' or extension.

Having emerged from the void, the solid line of the now prescribes and forms the entire spiral girdle of eternity in its own abstract primal space. It completes the girdle containing all the worlds and minds of existence in one instant. That instant is equivalent to time's own substantive mass: energy/matter. In that instant nothing becomes something; then becomes everything, and then disappears back into nothing — while I continue.

A JOURNEY THROUGH THE HIGHER GALLERIES

So far we have observed the making of eternity out of nothing. We will now observe the other phases of eternity leading to the creation of the universe and finally to the human psyche at the bottom of the spiral where the now disappears back into the nothing. In this next stage of our journey down through the girdle we will be looking through the galleries of the eternal, universal, stellar and solar minds, leading to the formation of the terrestrial ring more than half way down the spiral.

But first of all, why is it a spiral? — because the solid line of the now, emerging at absolute speed, does not follow a 'straight' trajectory. Instead it is drawn down and inward towards the contracting centre into which infinite mind is disappearing. This forces the line of original time, space and matter into a huge spiral trajectory, establishing the curve and spiral as the fundamental 'shape' of subsequent universal time, space and motion.

The first enormous spiral circuit completed by the now forms the initial gallery of eternity. On its outer or 'other' side, it is completely abstract and non-existent — the intellect. On 'this' side, the inner side, through which the endless echo of eternity reverberates, it is absolutely solid, impossible of existence and therefore completely unknowable to intelligence. Within the spiral circuit, where infinite mind has withdrawn, is eternal mind — endless consciousness.

By the time the first spiral gallery of eternal mind is complete, infinite mind has contracted to a point of infinite consciousness at the centre — and vanished. (Infinite consciousness, like infinite mind, cannot tolerate even the negative presence of eternity.) Between the centre where consciousness has disappeared and the surrounding girdle, power lines arise like innumerable spokes of fine wire to create universal mind within eternal mind. Simultaneously, due to the enormous concentration of power at the centre where consciousness has disappeared, the universal I starts to arise.

From the arising I, a backsurge of power sweeps out towards the girdle in the form of gravitation. Gravitation is ultimate velocity, the speed at which universal I communicates. Ultimate velocity or gravitation is so fast compared with the speed of light at which our perception travels that from our viewpoint it is already 'there' and everywhere in all directions at the same time. There is no interval in gravitation. But although gravitation (universal perception) has for us the quality of omnipresence and is thus far 'swifter' than light, it is still finite. Although ultimate, it is not absolute like the now. Therefore the I's perception takes time to cross the huge stretch of universal mind out towards the girdle; in spite of its enormous speed it falls behind absolute time. Also, when it does arrive, there is no world for I's perception to reflect on. The initial gallery is too abstract and the I's perception would pass straight through it into infinitude, making existence impossible. Without a world or gallery to reflect on, I cannot exist.

So where is the world to come from? How can a solid, reflective universe be created out of nothing in this infinitesimal interval? The answer lies in the absolute speed of original time.

In this minutest of intervals the now is completing a second circuit — around the central arising universal I. This second gallery, less abstract than the first, now crystallises in time at the wavelength of universal mind. (The now always crystallises at the wavelength of the mind it is enclosing.) But the universal I has missed observing the formation of its own gallery or world due to the relative slowness of its perception. That knowledge, now lost forever, is the universal I's unknowable past.

The second gallery, or world of universal I, is the stationary universe. Immediately the I reaches the girdle and perceives the 'stationary universe' the universe begins to move. It moves because the intelligence of the universal I is less than absolute. Its knowledge of reality, or all time, is incomplete due to the gap or barrier created by its unknowable past, the solid universe. Therefore it must hereafter share in the apparent universal process of motion or change towards the rectification of that ignorance — evolution.

Immediately the universal I's perception reaches this second gallery the first system of existence in eternity is complete: I, mind and universe. Simultaneously, the universe stops forming, because the I's perception, having reached or caught up with the now, the present, is no longer accumulating unknowable past and therefore no more solid universe can be created. So where does the solid matter of the now go now? It is being consumed by the new universal system (I, mind and matter) to sustain its own existence.

The universal I is a superb intelligence and because it is universal it is relatively selfless. Nevertheless its positional bias at the centre of the system lends it a certain subjectivity. This is sufficient to create the perception of 'this' side, whereas before there was only nothing, the completely abstract 'other' side.

The knowledge of this first part of creation is preserved for all time in the form of the energetic starry universe. But it can never be known so as to be understood by perception. This is because the world immediately perceived or occupied by any I — including this sense-perceived world that I, the reader, occupy — must always remain unknowable. Each world is completely unfathomable to the I and mind in phase with it, because in each case the world consists of eternal time (or echo) that no I has yet arisen to know. In short, I am always separate from the world I occupy. I can know about this world, or about parts of it, but I can never know it as it really is apart from myself. This is because the world I occupy, or perceive as being apart from myself, is not really my world at all. Yet I assume it is my world; and that is the most fundamental, time-consuming, self-perpetuating error of all time.

In man's case, his world is not the external world he perceives but the inner natural world he intuits through sensation and being. One can never perceive one's own world; one can only be one's own world in the same way as man is his sensation and not his perception or concept of the world. This state prevails in life after death when man actually experiences the death-world, either as his own emotional, feeling self or his own consciousness. The latter is the state that while alive he yearns for as the sharing quality of constant love, but which persistently evades him as long as he looks for it in the unknowable matter and objects of the external world, which is not his. Really, my world is my consciousness, out of which I, intelligence, arise in every case. My consciousness is before I am and is present when I am not. I and my world are implicit in my consciousness. This world on the surface of the earth which I, as man, assume to be mine must therefore remain the source of my confusion until I die to it in one way or another and realise the subtle infinitude or truth of my consciousness.

Continuing our journey through the galleries we will observe the formation of the next two, which enclose the stellar and solar minds.

As the now flows on it makes two more circuits beneath the solid universal gallery. These appear as two loops of universal space. Universal space — the reality behind our sensory space — is what is left when the solidity of the now is extracted to create universal existence.

The top solid gallery of the universe and the two spatial ones immediately below it together enclose the three minds of the universal system: the universal mind within the solid gallery of the universe; the stellar mind within the first gallery of universal space; and the solar mind within the second spatial gallery. Furthermore, as the now makes its first spatial circuit, the stellar I arises; and, as it makes its second spatial circuit, the solar I arises. These do not have a solid world of their own; they are a part of universal mind and the universe is their world. But in each case I am present in the mind as a distinct state of consciousness.

As the now completes its second spatial circuit, the newly arisen solar I emits several ideo-rings into the surrounding solar mind. An ideo-ring is a cosmic or divine idea expressed by a star, in this case our sun. One of the ideo-rings is the terrestrial I. It embodies the idea of the planet earth, not as an orb of solid matter as our senses present it, but as a ring of consciousness containing the life-principle of Man.

Release of the terrestrial ideo-ring into the solar mind starts another chain reaction. Because it is a pastless and therefore negative ideo-ring, it immediately reacts to the positive gravitational field of the solar mind. This induces a new field of dynamic tension around it and forms the terrestrial mind.

Like infinite consciousness before it, the negative (spiritual or abstract) consciousness of the ideo-ring cannot tolerate the positivity of terrestrial mind, so it too vanishes. In its place arises the terrestrial I.

The terrestrial I's perception starts speeding out towards the stellar universe at the speed of light. To do this it uses the medium of solar gravitation lines which at this stage are being incorporated into the terrestrial mind. Due to the finite quality of light, there is once again a delay as the I's perception travels across the terrestrial mind. In that interval the now forms another solid, smaller gallery beneath the two spatial stellar and solar circuits. This gallery is the lost or unknowable past of the terrestrial I. Immediately the I's perception reaches it, the gallery stops forming and exists as terrestrial or planetary matter — the original substance/energy of the earth.

However, this original earth matter is completely 'dead', inconceivably dense and lifeless and unlike anything perceivable in our external existence. Today, all the matter we perceive possesses past; can be described, analysed, named. Hence our external view of the earth and its matter is always of some form. But original matter has no form, only absolute density, because it has no past. Past is the track of life, and original earth-matter is utterly devoid of any vestige of life, but that is beyond our life-given comprehension.

Also, our view today is from the externalised surface of the earth, but in this original phase of eternity there is no surface, just as there is no life. Today, nowhere on the earth or within it, wherever we explore or dig into it, is there a place where life cannot be found to exist in some form or another. This is because life is now 'here'. It has actually penetrated from within and permeated the original, lifeless girdle of earth matter in the form of nature and the species.

OUR ULTIMATE DREAD

The terrestrial gallery of the spiral girdle is the foundation of terrestrial mind and the rock-bottom of our psyche. Its peculiar consistency, as I mentioned, is 'unknowable past', whereas the substance of the girdle itself might be called the 'non-past', founded in the original sound of eternity which continues for us today as the now that never ends.

This 'non-past' is the fundamental area of the terrestrial mind and is not accessible to knowledge, although there is a terrestrial memory that stretches back beyond time and the past of life on earth to the moment when the 'terrestrial I' arose in the mind. Its cosmic name is 'the vast memory' and man, the highest intelligence, can have conscious access to parts of it, right back to eternity, if that serves the purpose of consciousness and the man is sufficiently evolved. This 'vast memory' is deeper than 'the collective unconscious', which only started to form with the first spark of life on earth but then continuously gathered in the mind as instinctive experience. Out of the collective unconscious come the survival instincts of all the species and of ourselves. Everything above the fundamental level of the human psyche, from instinct to conscious memory, consists of past extending back to the start of life itself; and is knowable, or recognisable, given effort and inspiration.

Although the girdle is totally devoid of past and is dead or absent for us in sense (therefore literally non-sensical and unable to be known) it persists in being apprehensible at the selfconscious surface of the human psyche. This gives it a sinister and frightening quality. It exists but it should not exist according to sense-based knowledge. It is 'there' but instinctively we know it should not be. Its presence challenges the very validity of our perceived existence; and it is not a pleasant sensation. Even the most insensitive of us sooner or later apprehend it; we 'know' it is there. We all know how mysterious and elusive the unconscious is — and that is full of past — but the 'non-past' is far more disturbing because it represents the complete absence of consciousness itself, of 'I' the knower and the known. It represents the ultimate dread of all selfconscious life — extinction.

The non-past is sometimes sensed in dreams, in the imagination and in delirium. The effect is of terror, of monstrous, impossible demands being made on the individual. It is the rock on which sanity perishes and madness survives. For those who have met it, it can best be identified or remembered for its demands or its presentation of the impossible. Once experienced, it is a tangible horror never to be forgotten. Knowing can never enter it or recall what is in it; but although inaccessible to knowledge, the presence of the 'non-past' as a state in the mind is demonstrable not only in individual experience but also in that most reliable of all human records — myth. The non-past at the bottom of the human psyche is the darkness associated in Greek myth with the river Styx, a stream through the infernal 'stygian' darkness of the subterranean world of the dead.

This hellish portent of the non-past is responsible for man's dread of death. Yet, remarkably, the non-past has nothing to do with death, for death is part of the past. The non-past at the foundation of the psyche lies beyond the past, where it can never touch on life or death; it cannot touch us, except as a threat or fear. Death is merely the unknown, which is only tomorrow, and that is not frightening enough to account for the dread of death; otherwise we would all be running around hysterically morbid at our inescapable fate. Death is merely the other side of life; it is familiar, intimately known by every one of us in its benign essence. In fact we are actually seeing through death now and do not realise it. No one at the moment of dying fears death; most welcome it and to all it is the most natural thing in the world. The only sad thing at the moment of dying is that the living do not understand.

As I was saying, all conscious life apprehends the primal non-past as the ultimate horror. It is sensed to be the absolute end — whereas in fact it is the absolute beginning. It is the mind equivalent of matter out of which life emerged and out of which in turn intelligence evolved. Man's position in relation to the non-past is represented by the human body: intelligence, the head, balances on the body of life which is up to its neck in stygian matter. When man as much as moves his intelligent head to look down, or apprehends the primal matter out of which he came, he is petrified by his predicament. But death protects him; for if he could choose to go on living he would gradually fossilise and disappear back into matter. Failing to see death as his protector he is terrifyingly aware of being fractionally above the impinging primal darkness. Yet death compassionately sees to it that he cannot fall back into matter.

So man's dread of death is a case of mistaken identity. He mistakes death for the vulnerability he feels in the presence of that primal source from which, through the ceaseless now of life and death, he has struggled to emerge and be free since time began.

Nineteen

THE PLANETS

*By looking into the reality behind
the planets we illuminate their
original idea. And get closer
to our own enlightenment*

303

THE PLANETS

THE PLANETARY IDEO-RINGS

IDEO-RINGS ARE IDEAS expressed by a star, and as we saw in the last chapter, several were released by the solar I as the now encircles the solar mind. Ten or twelve of them encircle the solar I. Each represents the consciousness and governing principle of a planet, and together they form a single band of ultimate orbital consciousness, or spirit, within the solar mind.

Each planet or ideo-ring has a different ruling principle. The consciousness of all planets is identical but the principles vary. The ruling principles of all the planets combined influence in turn the ruling principle of each planet. This is effected through any life expression that may form on the planet.

The ideo-rings represent the real shape of the planets and their real significance relating to human immortality and the grand scheme behind the information provided by our short-lived narrow-visioned senses. As perceived by the senses the planets themselves are very limited, formal versions of the ideo-rings and in comparison are very uninformative. The senses only present us

305

with variations of the point or orb of a planet. The point, as we have seen, is the initial shape of all matter coming into existence. In other words, all cosmic bodies whether distant or near, take their shape (a dot or orb) from the original point and expansion of the now; and the retreat of eternal consciousness keeps them at that size.

From the point of view of the human psyche, the ideo-rings are pastless and abstract in character. They can only be intuited. This is not difficult; it is merely a matter of recognising the truth of them as they are described. Furthermore they are incredibly potent or real, consisting of pure idealised energy (spirit) and thus are divine principles or incontrovertible laws governing existence. Together the planetary ideo-rings permanently occupy the first, finest and most central region of the solar mind. There they form the radiant planetary world of spirit and life, the paradise above the Seventh Level of Mind, or our Seventh Heaven.

The ruling principle of the earth, or the divine idea in terrestrial consciousness, is actually the consciousness of Man, not man as we understand him but Man in his preconscious, pristine and august state where he is one being, a real and objective character/idea. 'Terrestrial I' is the principle of intelligence behind and within all life on earth — the one and only Man, the Lord and spirit of the earth. In terrestrial consciousness, the higher mind behind and supporting the human psyche, there is no earth as such, but a ring of consciousness symbolised by the earth's physical orbit circling the central solar I. This is the first terrestrial ideo-ring of Man or intelligence.

A second terrestrial ideo-ring emitted by the solar I glides out through the first and stops just outside it, adding a brilliant ring to the radiant band of terrestrial spirit. This is the ideo-ring of the moon, the lunar principle. Nature originates here. Nature is the life process of the earth; but it is not life itself. Life itself is cosmic and manifests on earth and to earth-man in the process and forms of nature.

This order of ideo-rings in the mind — intelligence surrounded by nature — is symbolised in the physical world by the moon, the nearest cosmic body, orbiting man and the earth. As Man represents the first terrestrial principle of intelligence so the

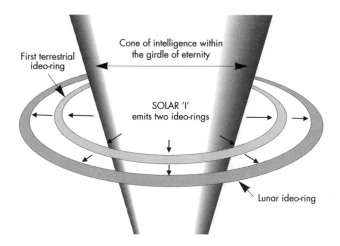

Terrestrial Ideo-rings

moon represents the second principle of natural or organic life. The lunar principle determines natural or organic life as a fluid or liquid process expressed in flow and rhythm. Essentially, it manifests as juices, acids, sap and the flow of blood.

The release of the pastless lunar ideo-ring into the terrestrial mind within the girdle of terrestrial past causes another sharp reaction. A new field of tension — vital force — charges the surface of the mind between the two. This sets the stage for the final phase of reality: the forming of the human psyche and the externalising of life on earth.

How man forms the universe

Every object in the sky has been placed there by man. He is forming the external universe. The real universe — beginning with the ideo-world — is deep within his unconscious and since time began for him, he has been slowly projecting it through his senses via the intellect. If it were possible to return to a point before man became selfconscious, most of the exterior universe would be seen to be missing.

The universe manifests in proportion to man's ability to perceive and define it. Yet, although it is being formed by him, it cannot be said to be his creation. It already has total reality within the radiant world of spirit. Nonetheless, man is under relentless pressure from the central core of intelligence to give the universe ever clearer sense-perceptive and rational existence, which is precisely what he is doing via his telescopes, rockets, films, space journeys, writings and imagination.

I will now demonstrate how man is forming the universe, using the example of the moon.

The moon is the first cosmic body out from the earth and the nearest object in the sky to man. Its proximity is of great significance — no accident or coincidence. The moon was the first cosmic body to illumine man's brain.

The story goes back to millions of years before man became selfconscious. In the Evolution chapter at the beginning of the book I described the rending of the veil of opaqueness, the psychic membrane that lies behind all animal eyes; and how this rending enabled sunlight to penetrate animal-man's unconscious, thus allowing him to fork off from the rest of the species on his lone, long journey towards becoming a selfconscious human being. But lunar enlightenment occurred even before this. In a number of sub-human primates living in the East a stunningly brilliant light suddenly blazed in their individual brains. The illumination appeared to them like a great light covering the screen of their until now complete inner darkness. They had no selfconscious existence and no awareness of the external world. At that time there was no external moon in the sky. The only moon was the real moon, the resplendent lunar ideo-ring of spirit infinitely deep in the unconscious. The 'sky' consisted of the inner darkness of the species which for just these few particular sub-humans was now almost nothing but light. The blaze of light 'blinded' them vitally within, rendering them inoperative for quite some time; to us this would be like being made unconscious.

Animal man could not perceive the external moon in the sky at the time of this lunar enlightenment, but his brain, his future intellectual instrument, was illuminated by the idea of it. This meant that the moon could now start to be projected sense-perceptively and seen with eyes. After a time the dazzlingly brilliant light in the sub-human awareness reduced in extent and power to the size, intensity and reality of the full moon. Eventually it faded in phases; and then in phases returned again to the full, lighting up the sub-human awareness over a period equal to our twenty-nine and a half days. Without physical eyes to see it, or the cognition to know it, the moon shone brightly in the sub-human awareness.

This is the same moon we see today. But our overt intelligence or selfconsciousness, which came a long time after, has seemingly exteriorised it, transferred it 'outside' as an object to be seen with the physical eyes in the same way as man has materialised himself as a sense-object outside his inner awareness. The moon still shines where it first shone, but while we are alive we cannot be aware of it there, because even when the external world disappears for us, as in sleep, we remain either subconscious in a dream-world or unconscious of anything at all. This is so until the day we die, when our awareness passes back through the sense-complex of the psychic brain and into what was the preconscious and is now the teeming psyche, where the moon still shines energetically and sky and space are filled with all the wonders of the natural world.

The moon was the first cosmic object to be intuited and then projected in sense by the human mind and we follow this same procedure today: the individual scientist, for instance, intuits the existence of a new sub-atomic particle and then discovers what he intuited. In other words, man forms his universe out of what is already in his brain. Nothing can be perceived or discovered in the external world that has not already been intuited or preconceived in this way. The intuition, however, is not individual; the man who intuits the existence of a particular phenomenon is not necessarily the one who actually discovers it. Frequently the same discovery is made simultaneously by different men in different places; and this is due to the cumulative intuitive vision of many men before.

An original or cosmic idea, such as the moon, does not enlighten the individual man, or self, but enlightens the psychic brain, or brain of the species. This is an especially important point. Individual man has to evolve to the point where he can realise the original ideas in his own personal brain and thus change its constitution and power. Until he does this, he is forced to use old cells in his brain that have been superseded by enlightenment in the psychic brain. In that respect man can be said to operate below his potential as a human being. When an individual man realises an original idea, or replica of an ideo-ring, it is rightly called realisation — the realisation of the illumination in his personal brain of an enlightenment which has already occurred in the brain of the species.

LIFE ON OTHER PLANETS

Since organic life on earth is determined exclusively by our moon, it is most unlikely that the nature of life on any other planet is the same as life here. However, the character of that life, as distinct from its nature, may be perceived in our own organic lives because all planetary life in the solar system is determined by the combined influence of the planetary ideo-rings, even events in our subconscious mind. Our day to day behaviour and motivation, as much as lucid dreams or visions, may be traced to our unconscious. But our unconscious goes so deep that it embraces the entire solar system and even the universe itself. This means that the characters of the ideo-rings of the other planets are influencing us. If Mars stands for activity through self-assertion, we have plenty of that. If the Venus ideo-ring symbolises union through harmony and cooperation, we now know where those impulses come from. Jupiter's urge for the expansion of justice, mercy and true knowledge is no stranger to us; but neither is the rigidity and caution of Saturn which makes us hesitate to always be fair, honest and true. And so all the planetary ideo-rings influence our lives and, together with the terrestrial ideo-ring, actually make up our lives under the dominant organic influence of the moon. Anything without organic properties is lifeless to us. So we are unable to recognise the planetary characters in their own right. They have to appear in an organic or lunar form; and that form of course is our own body. We are a composite expression of the whole planetary system. But our lack of self-knowledge prevents us from knowing this.

As the first principle of terrestrial life is intelligence manifesting through the idea of man, and the second principle is biological and organic nature appearing through the idea of lunar ebb and flow, so the third principle is character determined by the terrestrial ideo-ring and to a lesser extent by the other planets. So life on any of the other planets in the solar system will be influenced by the character of the earth and the nature of the moon, although the essential character of life there, whatever it is, will be determined by the specific principle or ideo-ring of that particular planet; and what that is in its own right we do not know.

It is almost certain that no other life like life on earth exists on any other planet or anywhere else in the universe. Life on earth is biological and organic — sensory. And sensory life depends mostly for its existence on the interaction between the earth and the moon. There are innumerable other cosmic factors too profound for anyone to detail, but some of the obvious ones that would have to be repeated to produce life as we know it on another planet are: the precise distance of the moon from the earth; the relative sizes of the moon and the earth; their orbital speeds, rotation and distance from the sun; the intrinsic power of the sun; the gravitational effects of the other solar planets

311

determined by their precise size and distance from the moon, earth and sun and each other, together with their rotation speeds and the eccentricity of their orbits. The odds against sensory life elsewhere are so great that finally it has to be said that the only place in the cosmos that meets all the criteria for the development of sensory life is exactly where we are. And because the cosmos we are considering is itself purely a reproduction of our sensory brain, the reality behind its appearance is in any case completely unknowable to us. Nevertheless, it also has to be said that life is life — whatever life is — and it does not have to take form where it is. It only takes form for us, through the interpretation of our sensory brain.

Twenty

THE MYTH
COMES TO LIFE

*In this last chapter we go back to
the beginning of life on earth, and
the final phase of reality expressed
through the girdle of eternity — the
formation of the human psyche*

313

THE MYTH COMES TO LIFE

THE FINAL PHASE of reality begins as the ideo-ring of Man or intelligence (the terrestrial I) emits a ringed replica of itself. This replica can be called the psychic I, the original empiric or evolutionary soul. Here begins in eternity what religious tradition calls 'the journey of the soul'. This is the original journey which all men on earth are making through life and death and whose final solution all must eventually emulate.

On release, the ringed replica of intelligence is pastless like its alter-self, the terrestrial I. It glides out through the radiant world of spirit (the terrestrial and lunar ideo-rings) and in passing through the lunar ring receives an abstract impression of lunar nature. It now has both the character of intelligence and the nature of the moon; but only potentially, because the replica still has no past. Immediately on leaving the world of spirit and on entering the vital field, the replica begins to gather both vital energy and past — psyche. Pushed outward by psycho-spiritual waves and pulses emanating from the world of spirit, the ringed

315

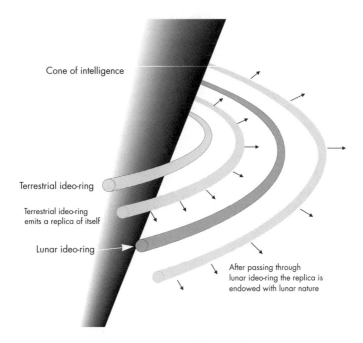

The emergence of the replica

replica slowly expands as it moves towards the girdle, every instant leaving its alter-self, the terrestrial I further behind. Past or psyche accumulates around and behind it in direct proportion to the distance it has travelled from its source.

At first, due to the absence of psyche or past, the replica's awareness registers no existence beyond the level of dreamless sleep. But proceeding deeper into the vital field it amasses sufficient past for reflection, sufficient psyche, to start to dream. These most obscure of dreams, including vague possibilities of future life, arise within the replica itself from the impression of nature that it received when it passed through the lunar ideo-ring. Mixed in with the dreams is the intrinsic urge, radiating from its

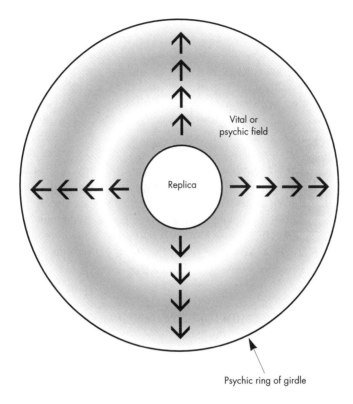

The journey of the replica

alter-self, to make life or the replica more conscious. Man today still has these distinct and vaguely prophetic dreams seeming to require the performance of impossible tasks.

Finally, after what in our time would be innumerable aeons, the replica comes to a stop; hard up against the terrestrial gallery in the spiral girdle. The replica has now gathered a fat body of psyche around its original ring of intelligence and is shaped like a puffy, ringed doughnut. Further, the whole area behind it, back to the world of spirit, now consists of its unknowable psychic past.

On the journey out across the vital field the replica received virtually no information from the 'waves' that were carrying it. But now, jammed hard up against the girdle, the replica is being bombarded with waves of information both psychic and spiritual. Within all this data and made very obscure by psychic distortion, is the ceaseless urge to pursue the task of intelligence — to make life or itself more conscious. From deep within its acquired psychic body, the replica's true identity of abstract intelligence (its empiric soul) starts responding to the call of the spirit. Between the pressure from within and the waves pounding it from without, the dreaming replica awakes to ego consciousness, the awareness of being I.

This egoic I, however, identifies only with the replica's acquired psychic body, its immediate position — not with its eternal state, the evolutionary soul. Trapped beneath the lifeless earth girdle, the replica can go neither back nor forward. It is like a pocket of air trying to escape through granite. Somehow the replica's organising ego representing life and intelligence has to penetrate the lifeless girdle of earth matter in front of it. Somehow it has to find a way out; or as with hindsight it appears to us today, to put on the surface vital life in the form of the species, and then selfconscious life in the form of creative man. No more formidable task ever confronted egoic intelligence.

THE EGOIC INTELLIGENCE REALISES ITSELF

Two simultaneous events now occur. The first involves the replica's psychic body. Pressed hard up against the terrestrial girdle, this is reverberating to the dynamic of the now coming through the girdle — the echo or word of eternity out of which all existence comes to mind. Under this influence, and at the

318

imperceptible pace of evolution, the psychic body goes through a metamorphosis. The outer region of it, touching the girdle, evolves into a great ear. This is the original ear of which all externalised ears are copies. Inside the ear a bony structure forms, resonating to the vibrations coming through it from the girdle. Deeper in again, and influenced by the developing intelligence of the ego as the ideo-waves sweep in from across the psyche, an embryo brain section starts to form. Slowly this organises itself into a chambered system for interpreting and evaluating the inner ear's reproduction of the eternal sound. This is the original psychic brain on which all subsequent physical brains are modelled.

The outer body of the replica is at this stage an ear and a brain; a complete, independent system for translating the echo of the now into the psychic frequency we experience today as sensation. But the brain's perceptions of reality have no life. The effect is one of complete mechanical detachment, as though the brain were a camera registering the images of reality without in any sense being a part of it. Also, at this stage, the whole brain/ear apparatus is purely a receiver.

Meanwhile, deeper within the replica's psychic body, near the point of its abstract consciousness, the other major evolutionary event is in progress. Here, at the root of the ego, far removed from the mechanical ear/brain function, great pressure is coming from the spiritual ideo-waves. These contain a solution to the replica's psychic entrapment; one which will further the task of intelligence to make life conscious. The ideo-waves are demanding and forcing an inward-going spiritual awakening that corresponds in magnitude to the psychic evolution of the ear and brain.

For a cosmic principle like life as we know it to start, a cosmic trigger is required. The cosmic trigger of life is the sun. But in this phase of eternity there is no sun, just as there is no external earth or world — because there is as yet no external life. For external life to exist requires the external sun; and for the external sun to exist requires external life. This is a deadlock that can be broken only by a spiritual or cosmic initiative in the mind. The egoic I must be made to realise the true sun within — the sire of the idea of the earth — by achieving a reunion with its alter-self, the 'terrestrial I' in the ideo-world. Its deeper awareness has to be induced to journey back to the radiant light across the psychic field, which is equivalent to its own unconscious. In a sense this means having to consciously undo all that has so far been psychically or psychologically accomplished. So, against the full onslaught of the psychic waves of past, and unknowingly guided by the spiritual pulses, the egoic I starts to dissociate psychically from the past and forces a line of consciousness back across the psyche to the world of spirit.

We must endeavour to appreciate how enormous this effort is, because any of us who venture to realise the truth will have to make the same journey with the same invincible persistence. The intelligence first has to disengage its habitual attention from the overriding importance that our brain (and the brains of others) attaches to the external sense-world. It literally has to roll back time, its unknowable past, and shatter the familiar, comforting psychic cocoon of dreams that has been spun around it ever since the ideo-world of spirit was left behind.

As the intelligence pursues this path of psychological self-dissolution it is assailed by near-irresistible physical and psychic illusions, blandishments and enticements to go on dreaming and to forget the effort of becoming radiantly conscious. But by sheer strength of character — the same admirable spirit, determination and restraint that man displays today in any noble endeavour — the intelligence dissolves its identification with its psychic body and past. When enough has been dissolved, the intelligence makes direct contact with its alter-self and the radiant inner world. In a brilliant burst of light the intelligence realises the true sun, the spirit of life. In the same instant the self (all identification with the past) is obliterated. In its place, shimmering phoenix-like with terrestrial conscious-ness, is the enlightened 'psychic I'.

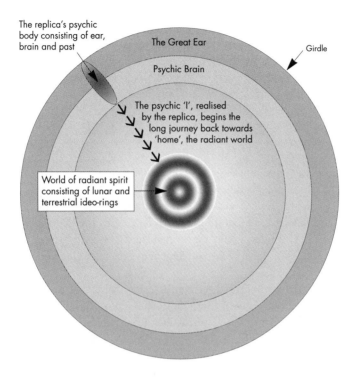

The replica's psychic body consisting of ear, brain and past

The Great Ear

Girdle

Psychic Brain

The psychic 'I', realised by the replica, begins the long journey back towards 'home', the radiant world

World of radiant spirit consisting of lunar and terrestrial ideo-rings

The journey of the psychic 'I'

Cosmic power from the radiant inner world streams through the psychic I into the outer brain together with the realised lunar principle of vital life. The cosmic connection — the trigger for life on earth — has been made. I have actualised the sun. Immediately, through the great ear pressed to the girdle, the brain begins to register sensations of the external sun and moon. Hazy perceptions of other 'outer' conditions also occur. But most significant of all, among the sensations being received through the girdle, are the first signifying life on earth.

321

LUNAR ENLIGHTENMENT

Released from its entrapment against the girdle, the replica of intelligence, the psychic I, successfully recrossed the vital field within the terrestrial gallery. Against all odds it reached and merged with the radiant lunar ideo-ring, the real, spiritual moon. The psychic brain 'listening' at the girdle was immediately enlightened from within with the complete idea of biological and organic life that the moon represented. This enabled the brain to begin the task of externalising more complex and variable life-forms on earth.

Later the replica crossed the remaining space and merged with the terrestrial ideo-ring. As this represents the life-principle of Man, the brain could then organise his externalisation and begin to realise the task of intelligence — to make life and itself more conscious.

The psychic brain is the governing brain of all the species, the nucleus at the centre of every individual physical brain on earth; the original of the ever-recurring individual reproductions. Being the one brain behind each man, it also contains the sense-complex that creates the sense-perceptive world man sees and evolves in. But lunar enlightenment of the psychic brain was not registered by all the species; only in the individual brains of the most evolved sub-human primates did the illumination occur. Lunar enlightenment produced the most radical changes in brain cells that had occurred to that date. But even so, those changes — like the enlightenment itself — were the climax of a long process stretching back several million years to when the terrestrial ideo-ring first emitted the replica of itself into the psychic field.

The first appearance of life on earth coincided with the passing of the replica out of and through the lunar ideo-ring. The further across the vital field the replica moved towards the girdle, the stronger was its influence and the more pronounced and complex became its effect in the form of the species. The replica's characteristics became the characteristics of the species

— a series of biological, hereditary facts that determined how our bodies from that point on would function and react, and even how we would think.

The approach of the lunar replica provided the irresistible pressure behind evolution. Its closing proximity forced continuous changes in the brains of the species and led to mutations corresponding ever more faithfully to lunar characteristics. When the returning replica merged with the lunar ideo-ring, shortly before lunar enlightenment, the idea of mammals was released. This produced warm-blooded creatures that suckle their young, and finally the primates — apes, monkeys and earliest man. Then lunar enlightenment raised the subhuman to incipient human status. It also had the effect of reconstituting old brain cells which had been in disuse since the early days of the species, and investing them with the higher functional and behaviourial lunar characteristics of man. This completed the irrevocable division between him and the rest of the primates — and started the final run-up to Homo sapiens.

The lunar influence on life was responsible for the female menses. Various female primates of breeding age menstruate every three and a half to five weeks; and it was the approach of the lunar replica which caused this. The subsequent lunar enlightenment of the human brain was responsible for woman's particular menstruation period, as well as for changes in the coupling habit and numerous other uniquely human idiosyncrasies and inclinations. Lunar enlightenment also engendered in man his uniquely sympathetic feeling-response, which would eventually manifest as the capacity to care and be concerned for others. This sympathetic response, as against purely instinctive response, prepared the way for the human memory and for imagination based on emotional recall, on reliving the past, and later for visualisation. From here on, man learned to feel, cherish and nurture beyond the hold of instinct. And as a compensating release for his new-found but more demanding humanity, he developed incipient compassion, and the power to sob and cry tears of pain.

And it would not be very long before he lost contact with his intelligence, forgot his origins and completely lost touch with reality.

Appendix

ORIGINS
OF THE BOOK

ORIGINS OF THE BOOK

IN READING THIS book you may have wondered where its knowledge came from. Who is the author and how did he come to write such a book?

Barry Long was born in Sydney, Australia in 1926. For the first thirty years of his life he showed no interest in anything profound. He left school at 15, went to work in newspapers and became a journalist, eventually becoming editor of the Sydney Sunday Mirror. But in the early 1960's he began to change and under the increasing pressure of a spiritual awakening he abandoned his former life. After a brief spell in India, where he went through a 'mystic death', he went to live in London. There, in 1968, he had the transcendental realisation which marked his entry into the realm of perception demonstrated in this book.

Through the 1970's he came to the attention of a number of people who became his first students and by 1983 he was teaching meditation and self-discovery at regular classes in London. In 1986 he returned to Australia to live and as the years passed he became ever more able to express the truth of life in a simple and direct way. He continued to teach in England and from 1990 to 1993 devoted himself to travelling the world. He has been able to speak to many thousands of people in Australia, Europe and America without having to create a large organisation or becoming a celebrity. He has been a profound influence in the lives of many men and women who seek to live a spiritual life. Through his own realisations, recorded in books like 'The Origins of Man and the Universe', he continues to make a significant contribution to us all.

The following appendix puts the book in the context of his life and work and answers the question: where did such knowledge come from? It is an edited account of a conversation between Barry Long and Clive Tempest during Barry Long's 1993 World Tour.

THE INSPIRATION

CT: Barry, I'd like to talk to you about how you came to write 'the origins book' and how you see its purpose in terms of your own life and work. So, first of all: what was your original starting point?

BL: Well, it actually began when a friend, David White, showed me a small book he'd written as an introduction to judo. The presentation was very simple and well done, with illustrations. David suggested that we write a similar book about the origins of all the martial arts — the power behind the force. It seemed so simple an idea. David was to write the practical part and I was to write the philosophical side.

CT: You had no personal experience of martial arts?

BL: No. But David was a Black Belt in judo. We also wanted to include the western self-defensive sports. Somehow boxing and wrestling didn't seem to fit into a work about fighting 'arts' but we decided to lump them in as best we could, along with other examples of violence — brawling, vandalism, terrorism, the lot. It was my job to find the thread that links them all together: to explain why man must fight at all. I suggested a title — 'The Fighting Spirit'. But how I would justify such a high-sounding title for a book about violence I didn't yet know.

It wasn't long before I was writing about the opposite of the fighting spirit — the fighting force. Quickly the whole thing just developed — until David said, 'Look, this is becoming a book in its own right. There's no purpose in trying to do what we started off with. Why don't you just keep writing?' So I did and the whole book unfolded. Without knowing it I had started to write a book about an aspect of man so vast that I shall be dealing with it until the day I die.

CT: This was sometime in 1978, while you were living in London and before you were recognised as a spiritual teacher. At that time you earned your living by writing astrological commentaries for a commercial publisher . . .

BL: Yes. I was teaching too — a small group of people who met in my house in London once a week — and I'd started to farm out some of my writing contracts, which gave me more time to devote to this book.

CT: Did it occupy you consistently from that moment?

BL: For about five years . . . I'd write and rewrite section after section, not in any particular order. I didn't begin at the beginning, but fairly early on I wrote the first section about science and religion, as an introduction, because that's the basic division in human thought.

CT: As you worked on the material you were going deeper and deeper into one subject after another. So although the book starts as a fairly straightforward account it gets more and more profound, drawing on increasing degrees of self-knowledge.

BL: There's so much in this book . . . it's like a base which my whole teaching goes back to, or extends from. My current teaching might be more refined, but is always consistent with it.

CT: The language might be different, but the knowledge is the same?

BL: Yes. The revelation to the mind or intellect which is

speaking, which is Barry Long, is a continuous process of living. And God-realisation is a continuous revelation to the intellect of that unfoldment. What I have knowledge of now is far more refined or more particular than when I wrote the book.

CT: Your teaching seems simpler too, because so much of it relates to practical day-to-day living. Whereas the book starts with remote events. What was it that took you back to the beginnings of time? Taking the chapter on the gods for example, how did it happen that you suddenly saw that there's more than mere legend to ancient myths?

BL: I say that the myths were not fictions, but represent the actual living experience of man and woman in that phase in evolution. The Greek gods displayed themselves, walked the earth. They were not Godmen or Godwomen as today I might say I am a Godman, or a master is a Godman. They had no knowledge or expressed appreciation of the one spirit or being behind their godhead. They awarded that to one of their number, such as Zeus, or took it upon themselves in the form of separate personalities — whereas the divine being is total absence of any independence from the whole; having no wanting or trying, no attributes, no need to even be.

Where we are now the gods have disappeared forever. You and I are these gods in a different time, extended into a totally different matter — because time is matter. The thing now is to be the gods in this different matter and time, so as to be a Godman or a Godwoman. That is to be completely surrendered to the most high, which is the most deep; to realise that profundity inside the body listening to these words or speaking them.

CT: Can you say how it happens that such knowledge arises in you? When you suddenly realised one day that you wanted to write about the gods of myth, were you just looking in the bathroom mirror?

BL: It happens now. All I have to do is look at a subject and the truth of it is revealed to my intellect. If you ask me to look at a leaf, or anything, and give you the truth behind it, I can do it. When I sit down to write anything, that's what happens. I couldn't sit down to write a novel. It would end up to be the truth of some aspect of life. There's no such thing as fiction where I am.

CT: And yet there's a story-telling thread running through the whole thing — 'The myth that came to life'.

BL: As the book says, the way to get as close as possible to the truth is through myth — not fiction, but mythic telling. For instance, I might tell the myth of man and woman, a most important part of my teaching . . . there are five thousand million bodies but there is only one man and one woman. That principle is endeavouring in each one of us to manifest in the senses, right up in the front of the brain; so that I am the closest possible expression of that divine man or woman that I can be in this matter now. In that I embody a great knowledge of truth.

That's not to say that I am divine. I am matter. What is speaking is force coming through matter. But its origin is another matter, we could say.

CT: What suggested all the different subjects in the book? Did questions just arise in you or were they prompted by life in some external way?

BL: Well, first of all we have to go back to how Barry Long suddenly changed at the age of thirty and started to speak the truth, or the beginning of the truth that I speak now. It amazed Barry Long that such truth could come out of his mouth. This paralleled an enormous change that had happened within. People talk about going 'within' but for me it was an actual reality. As I started to descend into that place there was more and more truth, as revelation. One of the astounding things was that wherever I went I would speak to the truth with such certainty that it was continually said to me, 'Well how do you know this? How can you say that? Everybody's entitled to their opinion.' And I'd say, 'Nobody is entitled to their opinion. There's only the truth and what I say is the truth.' Of course this was very extraordinary to them, especially coming out of Barry Long.

CT: Why 'especially'?

BL: Because I'd been such an 'ordinary' man. My aspirations, my work as a newspaper-man, my whole life was entirely based on intellectual materialism. I believed in science and 'cause and effect'. That world certainly exists but only in the frontal part of the brain. The inward-going gets rid of it. The state that Barry Long started to enter is deep within the brain. And as I descend into it the knowledge is enormous.

THE KNOWLEDGE

CT: *How would you make the distinction between just knowing something and contacting the knowledge deep within the brain?*

BL: As I descend into that state I am going towards what would be called 'gnosis', or in Indian philosophy 'jnāna'. Both words have the same root as 'knowledge'. But it is not what scientists call knowledge or the knowledge that anyone has as a result of their external experience.

Human knowledge is based on experiencing my place as an individual in the sense-perceived world, and my endeavours to connect various experiences up to make sense of them. It necessarily follows that there is cause and effect. I know in my experience that if I drop a glass it is likely to break on the floor. If I put my foot hard down on the accelerator when I start the car, it's likely to bound forward. That's what happens in most cases in my experience — which is always in the past.

What happened yesterday isn't necessarily going to happen the same way today — because the past is not reliable in the present. The newspapers are filled with what's new, which is what we didn't expect to happen. That's the uncertainty in our lives; the 'uncertainty principle' which science recognises. Nevertheless, all human knowledge is based on the probability that the past (the cause) will repeat itself in the present (the effect).

Human knowledge is based on the experience of an

individual mind or body. Real knowledge is not dependent on cause and effect or on the frontal part of the brain that we call the human mind. It is based on the being of the human race — that is, life on earth. Real knowledge is impersonal. Human knowledge is based on the person. To get human knowledge there always has to be a person. I have human knowledge, so there's a person in me; but I have no problem with that person.

CT: There has to be someone to know something. But if the something is not in the personal past or memory where does it come from?

BL: The real knowledge of truth draws upon what in the book I call 'the vast memory'.

CT: Is that the same as the thing you call 'the psychic brain'?

BL: The psychic brain is part of it — my past, my vast memory.

Whereas human knowledge has access to the experiential or impressionable memory, truth is the knowledge of the vast memory, which is a sort of repository . . .

CT: There's something in occult teachings called the Akashic Record . . .

BL: I don't use that term. Everything to do with the occult is a half-baked truth repeated from the past and never consistently original.

Another word for this vast memory is 'being', which can only be described mythically. I can get up and talk about anything — right back to the beginning of life on earth, in detail, from A to Z, which is now — but I tell it mythically. If I use human knowledge I've got to be very careful; because human knowledge invents more and more words. And the more words you get, the less truth. Although I use many words, you will find that I use them rightly and I demonstrate what particular words mean.

I say I am master of the West. Because the West is filled with words, you could say I am master of the word. That makes me 'Logos', a term applied to Jesus. As I am Logos of this time I have the power of the word and the power to use words. Not by any power that Barry Long is, but by the power that brought about my gnosis and is responsible for the real knowledge that appears through this intellect.

CT: Would you say that reading the book calls for an increasing degree of self-knowledge?

BL: The last part, where I introduce reality, is most difficult for someone who hasn't yet got the idea of what self-knowledge is. You can't perceive reality except through self-knowledge; and you have to have a very deep knowledge of the experience of life on earth as your own existence — which your self obscures. Self-knowledge is like a light put down through the gloom. Reality is under the gloom so you've got to put the light into the self first, as knowledge.

'Man, know thyself'. To know my self is to be able to perceive reality and to have gnosis or knowledge as I've defined it — knowledge of the vast memory, which is the memory of life itself. I don't suggest that I am the knowledge of all life, for that would be the place of God; but whatever I need to know to answer any question that I am asked, from this position in space and time, that knowledge is provided for me.

WRITING THE BOOK

CT: Would it be true to say that the book belongs to a particular period in your life, a time when you were able to reveal this material? You couldn't have written it at the age of thirty and you wouldn't write it now.

BL: Yes. I would say that was a seminal period in which I revealed my vast knowledge for the first time. I was able to write it down and record it.

You know, when I went to India [1964], I took with me a tape recorder because every time I opened my mouth the truth came out and I wanted to get it down. But in India I never found a guru, or anybody to talk to at any profound level, so the tape-machine was little used — whereas now we are often recording things, like this . . .

To answer the question, every moment of life contains boxes within boxes. There is only a certain time in which you can begin a project. You can either do it in that time or you can't. Sometimes it's not ready. You'll start a project and it just dwindles. This is because the will hasn't embraced it yet; it's just an idea. You can have the best idea in the world but it just can't be expressed. It's not the moment. When it's right it just comes out. And so it was with this book. I just pursued it, although it did take several years. And then more years to edit it and put things into an order that made sense.

CT: *And you'd be writing a few hundred words every day?*

BL: Oh yes. But I could only write for so long.

CT: *What would make you drop one subject and pick up another?*

BL: I can write up to about seven paragraphs but then I'm likely to say, 'Right. I'm going to do this again.' So I go deeper into the subject. Two thousand words later I've only covered what was originally the first three paragraphs. It's very difficult to get up to what was the seventh — because I've moved on into the subject, or off it.

When you're writing the truth, usually it just comes out and you give the best account of it first time. But I'm not one to only give an outline. I refine it and get more detail into it. It is an endeavour always to get more and more precision in the words, to describe life in the living experience of the readers and not their imagination; and to make the words powerful enough to represent a segment of their life. So that they can say, 'Ah, yes! That's the truth!' It is the Truth and the Word.

Often one of my sentences has several levels in it and that's why you can read different things in it at different times. At the first reading it registers in a certain part of you, where your self-knowledge is and you say, 'Yes, that sounds true to me.' As you descend into a greater profundity or stillness, you're likely to see something else — the next level down. So when you come back to it you say, 'Yes. That's true. I hadn't seen that before. That's amazing!' As you keep listening to the truth, you listen with different ears and read with different eyes.

CT: Some readers of the first edition said that they could only read up to a certain point before they had to put it down. But if they started again later, from the beginning, they found they could go on and read further.

BL: And some people will read it again without going further, because they haven't absorbed sufficient of it to go deeper.

CT: You describe the book as a descent, a journey into the psyche. And as you went on writing, you were realising more and more.

BL: Yes. In descending back towards the beginning and going through various strata, I realise whatever the significance of each strata is; and that's called an insight or a realisation. But because the consciousness moves so fast, it is a process of trying to get it together, grappling with it. So many ideas are bursting off the intellect and being received that you have to move fast enough to be able to hold them — because the truth moves so much faster than thought.

CT: How did you handle that as a writer?

BL: A word, a phrase or a sentence would jump out at me. Often it wouldn't fit in. I'd cut it out, put a red mark on the paper and put it in the drawer. Later I'd go back to it. But some things just could not be said.

CT: It's a question of catching it in the moment.

BL: Yes. You've got to be able to embrace the intellect and keep up with the speed of the perception so as to be it at that moment and be able to express it. It's not easy. The truth is new and fresh every moment and there's always new things being revealed. You're looking into something. You see so much. But you're trying to stay to a theme. Every theme leads off into a new vein of ore . . .

CT: Also a vein of awe.

BL: Oh yes! — the gold of truth that you're seeing. You see the theme and then, running off from the theme, are tributaries which if you look into them are rivers. And you can travel up any of them. That's the wonder of looking into the truth, into the void that I am. I can follow up anything, any moment; although if I'm not looking to find anything or not asked a question, I'm nothing. There's nothing there. Nothing arising.

CT: There came a point in the process of working on the book when you realised that all the fragments had to be pulled together. So can you tell me about that?

BL: Well, Peter Kingsley worked very hard with me on that. And he helped me to put some of it into words that would communicate. Peter was one of my students. I acknowledge the tremendous contribution that he made to the original manuscript, editing, arranging and preparing it for publication.

Peter would have realisations as a result of the work. I remember him saying that one of them was just like a ball falling down the stairs. He came back to me the next day and said he just went down and down and down, 'I saw just what you talked about, and I was right down there.'

Oh, there was a lot of coming and going, trying to put the book into order.

CT: Eventually it was all done and finished. And you found a publisher.

BL: A miracle! The book was so far outside the normal realm of accepted experience. As someone once said, 'You cannot venture to write a book that is more than twenty-five percent new.' And this book was eighty or ninety percent new. Unless you're dead, such books usually have to be published by yourself. Anyway, it got to the editor of a publisher well-known for esoteric books [Routledge and Kegan Paul]. She read it and apparently said, 'This book has to be published.' As I say, it was a miracle, brought about by God.

CT: That was the first edition. When it came out [1984], it seemed like a book ahead of its time.

BL: Still is. But it went all round the world and people read it and wrote to me and extraordinary realisations happened to some of those people.

REALISATIONS

CT: We need others to help us see the truth, don't we? We need reflection. What part did your students play in helping you to develop the book?

BL: Being the first concentric around me, they had a very distinct function. They allowed me to speak so that I could hear the truth I was descending into. Whenever I speak to the people around me I speak to a sort of a mirror. That's a very important function, and it's why we like to talk to people. Although mostly in our society people talk a lot of nonsense to each other, even when they tell their troubles to someone it should reveal the source of the problem. As a teacher developing a teaching, I had that first concentric to speak to, which is a privilege. And the reflection of the people still allows me to go deeper and deeper and to realise my own vast knowledge, by speaking it.

CT: What impact did writing the book have on your daily life, or on your work as a teacher at that time?

BL: What I discovered as I wrote the book I would often speak about, and then I would write with the energy of what happened at the meetings. My speaking always reveals the truth that is inside of me. After a meeting I would go to the typewriter and write down what I had seen. I have always endeavoured to record my realisation immediately, so that it's not lost. That's my job — to investigate the vast memory, to be it and go into the profundity of being, so that I can report back to the people and inform them of what I find. For I am the life of the people.

That is how I came to the seventh level of mind. I must have been talking about it at a meeting so I went and started to write it down. That's when I discovered that I am surrogate for my fellow man and woman.

I'm at the typewriter and while I'm typing I have the knowledge that I'm in the seventh level of mind and am looking into it. I see that I am actually in the seventh level of mind and that this is an extraordinary place. I'm aware of great profundity and a great beauty; an unimaginable beauty. And then I

see a figure. It's not an apparition. I'm looking into myself.
There it is. It's vast, a tremendous figure meditating, as you'd see
one of those great buddhas in Indian temples, but vaster, an
extraordinary, massive thing. If you were using external vision it
would have reached up to the heavens. And I thought, 'My God!
This must be the Earth Spirit.' I knew I was in some extraordinary
place. And yet as I looked at this figure, there was something
wrong. It didn't gel with the sensation of where I was. I looked
under it and it was sitting on a rudely constructed chair, as
though some pieces of old wood had been nailed together. This
enormous figure, sitting on a box! Then I realised, 'My God . . .
this is the Sleeping Giant! This is the mental configuration of
God or holiness that has been erected by man's notions and
imaginings and his religious strivings through all his mental
effort. This is it, preserved forever. Every bit of ignorance in man's
mind is in this Sleeping Giant, looking so holy, and meditating
like a buddha.

And then I was suddenly in the place behind the Sleeping
Giant, if you can call it that. I'd seen through it, so I was in the
other place. And that is the seventh level of mind that I describe
in the book. The place of paradise, where there is nothing, of
course; not a single thing is there. It is absolutely nothing, but
energetically it is everything that the earth consists of in all its
beauty and wonder, everything that we love. That's the Earth
Spirit.

So the Sleeping Giant is like the guardian of the threshold of
the Earth Spirit. It's the utter, complete ignorance of religions and
man's mental world.

Man's existence has several veneers over it. The first is the
actuality of the senses that produces the externality of the earth,
the trees and the clouds. That's natural — made by the senses.
But over that is a screen of man's mentality through which he
awards significance to the natural or the real — which has no
significance except that it is life. Significance is like a finger
pointing away from the reality. 'What's the meaning of this?' Do
you need the meaning of a sweet breeze or the perfume of a

flower? But man goes looking for significance and projects himself into his mentality — a mental body six inches or a foot outside his physical body. He lives in that mental projection. To get back to the truth you have to come back to your physical body. That's obvious — anybody outside the body is said to be insane.

CT: In a way the book is itself a part of the Sleeping Giant, a construct.

BL: It is only a myth, remember. It's not the truth; but it is close to the truth. The book is complex but not complicated. Mental constructs of what life is about are complicated, and the area addressed in the book has been complicated by man's notions. The book is actually a de-complication. It's producing more nothing. It's a simple book if only you are still and look at what it says. You can use it to go into your own experience within. That requires a very fine focus on your own experience but if you can do it then 'the origins book' becomes a book of self-knowledge.

CT: I'm still looking for the significance of the book in your life . . .

BL: Well, I'd say it was the means of getting all that I had seen and realised in my life into one stream of truth. By this time, 1978 and 79, many things had happened to me. I'd been through my transcendental realisation ten years earlier. I'd been through so many things that my profundity was enormous. But from the time in India when I started to write the beginnings of my truth, and then in London when I was writing poetry, what I was writing was fragmented. It was all one teaching — even now my teaching is just an expansion of those writings; different, but the same truth — but I hadn't brought it together into one stream of expression. So 'the origins book' provided a focus or funnel for it. It was a gathering together of everything that had happened since Barry Long started to change and this truth, this gnosis, started to come out of him.

CT: Were there any other major realisations that arose directly through the writing of the book?

BL: Yes. There were several. There was the one where I suddenly realised the serpent; or I was a serpent, and it extended right over my forehead like a cobra. And from that somehow or other came the knowledge of Draco. I'd never read much about Draco, you know. But it was revealed to me that Draco is the point of this existence.

About the same time it was revealed to me that I am making a constellation that will one day appear in the dream of existence as a number of stars in the heavens. Don't ask me how. Everything's a dream out here. But it is possible to record truth, and that's how constellations are made. I should say that all truth is recorded in the stars. (That's the basis of astrology.) I say that I am making a constellation, or realising one that's already there in the divine mind, and it will appear in the heavens as an actuality. You know, I once started off a poem with the words 'you are a star' and since then I have actually been making or affirming stars. That's what my life is about.

CT: What was the connection between the constellation Draco and the serpent that you saw extending over the forehead?

BL: The serpent is a finer version of the dragon. In our culture we know of the mythical dragons breathing fire, like the dragon of St George. Really that dragon was man's sexuality. Chivalry was his attempt to overcome it. The serpent is a form of the dragon, but an octave higher, where love and sexuality have united in wisdom or truth. And that's what I realised. I'm not saying I'm anything special. I'm only a reflection of our times. 'The times' is the total self of man on earth and we are only little cells of that self. Now, in our times, it is possible to overcome the dragon of sexuality, not as the Christian saints tried to do, by fighting or suppressing it, but through the higher octave of wisdom or truth, supported by the body of love. There cannot be any truth without love.

That realisation led me to the cosmic dragon — Draco. The cosmic dragon is the enormous power of what we know as sex, love and truth — the combination of them. It's the most powerful constellation in our immediate psyche.

341

CT: It seems that one realisation just follows on from another and arises as you apply yourself to the truth.

BL: Yes. But where I come from, I can only realise what some other man has realised somewhere in the great life of the earth. I build on it in some way, with a deeper realisation, because we build on each other.

Around the same time I entered the Body Diamond or Bodhidharma realisation. (I described this as 'truly a state where nothing dies in the fullness of love . . . the understanding that this is the body of my fellow man, behind the brilliant diamond point that only love of him can bear, or dare, to enter'.) I didn't know what Bodhidharma meant. I actually thought it might refer to a 'body'. Then I was told that there was a guy called Bodhidharma, the patriarch who took Buddhism to Japan. It was his realisation and I had realised what he realised. Whoever Bodhidharma was, and whatever his philosophy, that's where this realisation came from.

CT: None of the knowledge in the book came from any reading, did it?

BL: Oh no. One of the things I had to do, very early in the piece, was give up reading any books or other masters. When I started to look into books I saw that I would get information from them and my mind would build on it. So I gave it up. I didn't want to go into knowledge that was already reported. I will read myth. But I won't read other people's opinions or anything of that nature.

CT: Was anything suggested by other people?

BL: Probably, because we're influenced by all sorts of things, aren't we? Certainly if someone mentions a certain word, or I read a sentence in a book, I am immediately able to go down through it into its meaning. So yes, I would have received information from various sources. But I can't remember reading anything that could give me the profundity of insight that 'the origins book' reveals.

CT: And what did you feel when the material was being revealed to you? When you caught hold of something, what was the sensation?

342

BL: In the beginning it's amazing when you start to hear the truth coming out of your own mouth. But then as you become consolidated in the truth as your very life, the sense of it being separate from me diminishes and the duality no longer exists. The amazement never disappears, for God is always amazing, but as the realisation of God becomes deeper and deeper and the truth suffuses through your psyche the duality disappears. You get into what a scientist would call the 'singularity' — the absence or the void of truth. In those moments I say I am nothing. And anything I say, I just say, like now.

GNOSIS

CT: Do you recall at what point you first heard the word 'gnosis'?
BL: That must have occurred very early in my awakening. Probably around 1960 to 1962. I was Press Secretary to the Leader of the Parliamentary Liberal Party [in New South Wales] and I had access to the Library of Parliament. I was still reading books in those days. The librarian was interested in religion and psychic matters and for two years I had sufficient time to read my way through most of the religions of the world, contemporary books as well as old traditions, much of which I've forgotten. But I must have absorbed lots of things.
CT: Early on in your book you refer to the original gnostic masters . . .
BL: Oh, that came purely out of my knowledge. When I started the book, writing about martial arts, I was writing about something that comes out of the East — the original East. The West is represented by the gun. You stand off from a man, shoot him from a hundred yards or in the back and he doesn't know anything about it. There is no art in it. Whereas martial art is to face the man — to face your self in the man — and use the power behind your self to confront and overcome him, not with force but with power. That is something from the East. I'm not

talking about the East that exists today. The East, where I come from, is the total life. It's not a matter of doing some oriental martial art and saying 'I'm a master'. The question is, 'Master: can you love a woman, the hardest thing on earth to love?' — because love is power and it's used in the total life, raising children and at the kitchen sink. To me, that power is the real East. A man who is indeed a master of his art will hear what I am saying. All art comes down to power.

CT: *Would you say the power of the book is its gnostic quality?*

BL: Yes. That quality is there because I started with what power is. I'm talking about the power embodied in man, not man's force.

The idea is that you have to defend yourself against the forces of the world, for existence consists of forces that are always encroaching on the power. They can't take the power but they can invade it when it's distracted. In the beginning of time gnosis was the power. This is the power of nothing and the master was its receptacle. He only had to be it. It rippled out from the centre and in that way he held back the forces of the world as embodied in ignorant man who came to invade the sanctity of the centre. There was no need of any physical force whatever. But the forces grew in man's imagination; he started to think, became more corrupt and therefore more forceful, more material and more able to use matter to invade the existence of the power.

Every master needs the support of power from other bodies. So every master you will find is surrounded by various concentrics, supporting embodiments of power; not of the same octave as he, but acting as a ring around him. They take the force for him. His power, like a pebble dropped in water, ripples out through the concentrics.

CT: *When you speak of the early period of man, you are talking mythically and not historically, aren't you?*

BL: Well, what is our history? How far do we go back? Three thousand years of recorded history; but that's not very far. I'm talking about ten, twelve, twenty thousand years.

CT: That's prehistory. What's the difference between myth and prehistory?

BL: Myth is the handwriting of our past. Prehistoric means 'before being written down'. All history is mental. Every place apparently has some record of a flood that wiped out the human race. Not true. That's all mental. I'm talking about the real story of the human race, when human bodies were psychic, only just coming into expression as a sensory, substantive apparatus.

CT: For many people the words 'gnostic' or 'gnosis' would be associated with certain sects of historical times, heretical groups on the fringe of the early Christian Church, whose knowledge appears in the Dead Sea Scrolls and suchlike. Can you comment on the gnostic aspect of religion?

BL: That's a degradation of gnosis. No gnostic master would ever embrace a religion, or a thing like Christianity — because it is already degraded. Gnosis is the knowledge of union with God. So why would you embrace something else?

What happened was that the followers of Jesus, those who laid down what Christianity was, came into collision with the last of the gnostics, who insisted they must be able to speak the truth now, without reference to anything in the past. That is the gnostic principle. But the Church Fathers said, 'No. You must have faith in Jesus. Only Jesus knows the truth. You can only speak through that.' So the gnostics could not hold their original place. They could no longer be the power, because they had embraced a religion. They had taken a man as God. But no man is God. There is only God; the void, the nothing. So the gnostics at that time were finished and gnosis was about to go underground.

CT: Do you mean that a gnosis or gnostic method of teaching has survived throughout historical times?

BL: It has to have been here. Gnosis has always been here; because gnosis is the knowledge of union with God. In the beginning man was filled with God, or one with God. There was no distinction, except as the most high. But when man started to

345

use his brain to think he lost that vision. He forgot that he was nothing but God in existence.

CT: Does gnosis require that it is imparted to others?

BL: Yes; you will always find that gnosis endeavours to get the human psyche, the people, back to the beginning where gnosis is, to that realisation of God or truth, that oneness.

I say I am gnosis in these times — in this distance into time. I say: I am not born, I do not die, and I am always here. Therefore wherever you find me I will be endeavouring to inform others of the way back, to take human perception back to where it perceives God. I am gnosis because I take everything from man's mind. I take his belief in religions. If he thinks he loves his children, I take the thought. I take everything from him because everything he's got in his mind represents his ignorance.

CT: It could be said that 'the origins book' puts a great deal into the mind — new ways of looking at things — and in a sense it's a 'scientific' book. You give us various 'models' to test against our experience.

BL: Where I come from, science is one hundred and eighty degrees different in direction to gnosis. Science takes me 'out', away from home. Gnosis takes me home. Science discovers more and more things and puts them into the mind; and it takes man and woman out into a non-existent universe of analysis and interpretation. But gnosis is a process of negation. You will always find that gnosis takes away the complicated concepts that have been put into man by things like science and all religions — because they always complicate the truth.

'The origins book' opens up this way of looking at things — so that there's more space. Gradually I take away even the bits of matter holding the space together. Eventually you have to say, 'My God! I'm nothing. I've got nothing.' And I say, 'Right. Now you are looking at the truth which is God. If it leaves you with anything, it's not God. For God is nothing at the beginning of time and I'm going to take away all the time you've absorbed, all the complication, everything your mind has thought about.' That's what I shall do for every scientist on earth who is ready.

It is a very simple matter. When the scientist's child is dying, what does Einstein's theory mean to him? Does he run to the blackboard and write the formula for energy? Science, in its search for the truth, is as absurd as that. It cannot live life as it is — only as it is on the blackboard or in the mind. I am gnosis so I put my finger on the ignorance of science. For there's no truth whatever in how science sees the universe.

CT: And yet you seem to delight in scientific models and methods.

BL: Yes. I can delight in the scientific approach because the scientist is man in the world looking at the fact. I can use the parallel of science to get man to look at the fact without getting emotional. If I talk about angels or Jesus or Muhammad, what happens? Man can't get to the fact because he's too busy protesting from his emotion. You never find the truth unless you can see the fact without emotion.

Science is unemotional but it is still ignorance. The error of science is that it draws conclusions from the fact. That is ignorance. Science is the highest form of ignorance in our time. But I can only reveal the truth in the midst of ignorance.

The scientist is never going to find the truth looking from fact to fact and drawing conclusions. First fact first. He has to look at the fact of his own existence. He avoids that.

CT: That's because he believes he should be impersonal in his scientific work.

BL: He believes that, yes. But he doesn't believe he should be impersonal in his personal life.

CT: If the scientist is always finding more things to question, he can't really be interested in a final answer. How do you see this dilemma?

BL: He's always in dualism. Yes and no, right and wrong, matter and space. He's trapped there forever and will never get around that difficulty. He will always be just a split second away from the beginning of the universe. Or he'll think he is. If I'm diving off a cliff and there's a chance of hanging on to a rope, what's the good of just missing it by an inch or a split-second?

All ignorance is ignorance and falls short of the truth.

I take you back to the beginning. What's the beginning in quantum mechanics? Or the study of the planets? (One is the same as the other.) I'll tell you. It's a singularity. What's the end of quantum physics? Where does it fall down? At the same point as the theories about the origins of the universe — just short of the singularity. No scientist can pursue it any further, because singularity is obviously where dualism ends. 'Singularity' is the scientist's word for the nothing out of which everything came, or the mind of God.

I must use ignorance to demonstrate the truth. So I use scientific knowledge, which the people accept as the most important knowledge on earth. Then I go further than science can — beyond the Big Bang — to the real knowledge of the beginning of the universe. Either science is the truth, or I am.

CT: *That sort of statement just isn't acceptable in science, where nothing is true until proven to others.*

BL: Truth can only be seen by the individual. It can't be seen by the masses. That's why, no matter how many individuals see UFOs, the masses will never see the truth of them.

The question for anyone is: can you see the truth in your own experience? It is demonstrated in your life if you have the clarity to see it. I can demonstrate how whatever is within your life is affecting your external circumstances. 'As within, so without.' We've all heard that great occult saying, which so confuses the people. By the grace of God I have the ability to use what is without to demonstrate what is within. I do this in the book. The Seven Levels of Mind, for instance, can be demonstrated in everybody's experience. I demonstrate them first in the reader's experience of his own mind, then of his own self, and then of his own existence. But the question remains: can the reader see his own experience? That's the hardest thing.

To look straight inside yourself needs clarity. People can't see straight because of the unhappiness, doubt, self-judgment, fear, jealousy or resentment that clouds the great, wonderful clarity called intelligence, personified in whatever is looking out of anyone's eyes at this moment. The cloud of emotion makes the perception bend, just as a stick appears bent when put into a river. Water is a representation of emotion. When you look through a medium like that you just can't see straight. I am not divided by emotion or my self. I use clarity to see straight to the fact and, through the fact, to the truth. And that is due to the grace of God, not to anything that I the speaker am.

THE POINT

CT: Can we turn to the final chapters of the book now, describing the structure of reality, and look at the key to it, which is the Spiral Girdle. Can you say how you came to that knowledge?

BL: I can't at this moment recall how I got it, but it's based on the very simple reality of how it all begins. Everything in existence begins with a point. You put your pencil on the page to write: it begins with a point. When the typewriter hits the paper there's a point where each letter begins. A seed is a point and if there is such a thing as an atom, it's a point.

I can see this with extraordinary clarity. The question is: what happens to the point? And the answer is: it immediately expands and if there were no restraint it would fill the whole universe. So there has to be restraint. That is what I realised in the Transcendental Realisation [1968]. That is the realisation that allowed me to write about reality.

The tree over there started with a point, a seed. It grows within its due bounds. The size of the leaf is the size common to that kind of tree, as is the size and shape of the tree itself. The tree has not expanded until it is as big as the universe. It has its particular form and it is held in that shape. Now what is the restraining influence on all the 'points' that keeps them in their due bounds?

In my transcendental realisation I realised that I was producing absolutely nothing. And that was because I was moving at such speed. To demonstrate what this means I have to use the example of the cosmos. If I move at speed away from the earth, it gets smaller. If the earth is coming towards me, for instance at twenty-five thousand miles an hour, but I am moving at the same speed, the earth stays the same size. If I move at only twenty-three thousand miles an hour, the earth will start to get bigger until such time as it overwhelms me, or I land on it.

It is the way of things in my consciousness that I am 'moving away' from all things at a speed that keeps them within bounds. If the tree were left alone without me, it would grow and grow but I'm moving away from it at just the speed that keeps it within due bounds. The hair on your dog's back is just so long, according to the species, and it doesn't grow any longer. It's the speed of consciousness that keeps that dog's hair at that particular length and stops it from filling the entire universe.

You have to move very fast to get the idea of what I'm saying. You can't necessarily realise it, but you can get the idea because it is the truth. You mustn't try to understand it; you've just got to receive the idea.

That is the state of my consciousness. It moves at such extraordinary speed that I am able to speak the truth as I do. That is an amazing realisation and I have never heard of anyone else on earth who has been able to describe it.

CT: It seems to underpin everything that is in the book, and in a way explains everything in it.

BL: That is so. This extraordinary consciousness, this extraordinary being, is the beginning of existence, the beginning of the universe. Therefore it is the beginning of me. Everything is beneath me. Everything is completely dependent on the speed of my perception, or my being.

CT: And that speed of consciousness means you are able to perceive anything as just what it is, as the idea of what it is. But what is an idea?

BL: An idea is that which is held in due bounds by my speed. Now, of course, the question is: where does the idea come from? Well, as the book says, there is a great silence and stillness, the void, and then suddenly an extraordinary sound comes across the intellect. A great 'crack'. It enters existence as a point.

What happens is that 'me' looks at 'my self'. Anything that exists is my self: 'me' does not exist. My self is now separate from me. So I'm looking at my self. The realisation is that everything I am looking at, the earth, the world and everyone, is my self — not the notion of 'Self' that we've got from eastern masters but the self as my ignorance. This whole blessed, beautiful, sensory perception of the earth is in fact my ignorance. And 'me', who's behind it, is God. Or nothing.

I am nothing in particular. I am neither greater nor more special than you. But somehow or other I am appointed, or commissioned to go into these deep levels of perception and then to come back out into existence and say: 'This is how it is. I have seen this.' But I am only surrogate for you.

CT: You were talking about where ideas come from. Pure idea is outside existence and does not come from self, so can you say how ideas come into existence?

BL: Yes. Every idea has to come into existence, once it's embraced by the will. So the idea is held within somewhere — within 'me', that realm of the unknown or what we would call God. Yet in that realm of the unknown is something that we can say is 'known' — because I say it contains ideas.

The earth, for instance, is an idea. The earth has been projected from inside of me out into a living idea, which is a condition or an object. The earth was an idea inside of me, which has been projected in sense. But it is only in sense. As soon as I go to sleep at night and withdraw from my senses, the whole earth disappears. Then when I wake, there it is: a magnificent idea called the earth, projected in sense so that I, the perceiver, can enjoy it. That's a divine idea. I who am speaking certainly didn't get the idea with my limited vision.

CT: It seems that you endeavour to seek out and describe the

divine idea behind each thing. Every cup is of the nature of 'cup'.
That theme runs through the book.

BL: That is so. And I'd have to say that this is where my descent into me passed through Plato's vision of things. The idea of that idea was expressed by Plato, who appears to have been the first to say it, although he would have got it from somewhere else and built on it, just as I do. I descended into the stratum where that idea is.

CT: Could we say that 'the origins book', and indeed your realisation itself, is the manifestation of a divine idea?

BL: I, man, have to manifest the idea and bring it into sense, just as the earth has been brought into sense. It's not easy, as everybody knows from their life. We're all trying to shape an idea and to manifest an idea, every one of us.

There was a part of my transcendental realisation where I was utterly and completely astounded by the attempt of sense to reach or describe my being. It's like looking down from a great height. I am the being. And there's this sense reaching up, trying to reach up to describe my being, but it's a thousand miles away and it can't just get there. It's trying to communicate with me, but it's utterly impossible. It's just so ridiculous, a big laugh, for I am so far beyond it.

The whole of my life since that time has been the manifestation of that idea. My teaching is all the ideas inside my psyche and they are still manifesting.

Going back further, to my awakening in 1960, when I first experienced what I would call the Lord, I realised the idea of God as Woman. That is fundamental to my teaching but only in recent years have I been manifesting it in actuality, in man and woman's body. Once it was only an idea. It was the Lord appearing to me in the vision of a woman. A woman appearing as God — a woman of all things! And I'm not talking about Kali, or one of those Godwomen ready-made by man's mentality or some tradition. I'm talking about an ordinary woman.

That idea had to be manifested. I, man, had to bring it into sense. And now I teach women to be Woman.

352

FOR MAN AND WOMAN

CT: You were not teaching women when you wrote the book.

BL: At that stage I had not developed the idea of woman sufficiently. I hadn't gone deep enough into woman for the idea yet to manifest. I have to shape the things that I'm bringing in, in actual sense. There are forces against it and forces for it. And then there's the power within. I have to get rid of the force and bring the power into the matter. It's not easy.

I could not touch too much on love in 'the origins book' but I can talk about it here. Where I come from, the idea of love is already manifest and it is man and woman's existence. Now I can address the love of man and woman, which is the same as the love of the earth and the love of God. And I address it truthfully, not according to the half-baked ideas of love that are inside everybody. The only love is that which is manifested in life itself. You could say man and woman's love is obviously manifest for men and women in existence; but, looking out at it, they get their own ideas inside themselves of what love is about. The truth lies behind the fact. They must look at the fact of love outside them, to find the truth of love inside — to get their love right.

CT: The book might seem to be more for men than women. After all it's called The Origins of 'Man' and the Universe . . .

BL: Man, the principle, includes man and woman . . .

CT: Nevertheless, you do address the whole story from the male point of view. Is that because it is man's myth and not woman's?

BL: It is man's myth but it is also for woman. Woman is love. I'll tell you the myth of woman's love. If you go back into gnostic writings, back to Sophia, (the divine principle of woman) you will find the tears of woman. She wept and her tears created the unhappy world. Why? Because she left the Father. In the supercelestial world of the unbegotten Father, she saw his androgynous image, immortal Man and loved him.

The Father said, 'You'll forget me, my love.'

'Oh no, I won't, Father! Never, never will I leave you or forget you!'

'But my love, you are already leaving me!'

'No, Father, I won't I tell you!'

So because she was distracted in her love she descended from that place of integration into the psyche. And as she came down she forgot the Father, her integrated state, and her love of the immortal Man. She took form in the physical, separate from both and started to love physical man. Being separate from the Father and the knowledge of immortal Man, she will always be in a vale of tears. For that's what it means to be separate from my love, either as man or woman — as everybody knows, for everybody is endeavouring to get back to real love.

Where I come from, it is man's job to actually be the living Father, the living God who says, 'My love, I am here in front of you in sense. I have entered this world to unite you once again with me. I'm going to show you that I love you and I will bring you back to me, back into me. Then we will not be what we are now. We will be back with the most high, you and I together, forming that great consciousness.' That is man's task. That is the whole basis of my teaching. That is what I am living, what I am doing and what I am teaching.

The most high is the absence of this sensory existence. That means absence of man and woman as physical creatures, absence of the psychic principles that man and woman are when they are divided in the psyche. When man and woman disappear, me and my love are one. I disappear and she disappears into that state of absence called God consciousness — just a state of being in which I am absolutely irrelevant because I am united with my love.

CT: To sum up, how would you describe the purpose of the book now, in relation to your current teaching?

BL: It is a journey to take you into the place that I now call 'me', the immediate presence in everybody of their own being. Everybody can experience me now, as that most intimate sensation and knowledge of life inside the body, before life takes form. It can commonly be experienced as the sensation of joy or wellbeing.

This being is as infinitely deep as the space we see in the universe, which is actually a re-presentation of the depth of me. As I descend into it, I start to realise more and more of God, truth and love. So the book is a means of descending step by step into a place where ordinary information and ordinary thinking are dispensed with. Here I have implicit knowledge without the need to analyse, categorise, or think about anything. It is the place of gnosis.

CT: *The final question: you talk about 'where I come from'. Where is that?*

BL: From being. I come out of the joy or wellbeing in every body. As I only exist where there is an object, the realm of 'I' is out here in existence, where all objects are. Wherever there are objects there's the subject — 'I'. But when I look inside, into me, which is a dark, endless, black place, I see nothing. For there is nothing in me. And if I look long enough at nothing inside of me, then I, the subject, disappear. At some point in everybody's life this disappearance into nothing becomes an object of terror. In the mystical or divine life I have to go through that. It is what's called 'the mystic death'. What happens when I disappear? Lo and behold — there is simply being. No subject and no object. No duality. Only the indescribable state which, for want of another word, is called being. As there is no object in being you could say there is no purpose in being. It is therefore an effortless state. For all effort is a struggle towards some aim, object or purpose. Being is effortless, because it is now. And that's the end of it. Nothing more can be said.

Peter's Point, Jamaica
8-9 June 1993

INDEX

OTHER BOOKS BY BARRY LONG

THE WAY IN – A BOOK OF SELF-DISCOVERY
A revelatory book, THE WAY IN is uncompromising and profoundly powerful in delivering the truth. With every sentence you penetrate to a deeper level of self-discovery, and an ever wider vision and freedom.

KNOWING YOURSELF
By being exposed to the many false layers of yourself, you come to self-knowledge. A classic statement of spiritual realisation.

MEDITATION A FOUNDATION COURSE
A proven bestseller that gives you simple but powerful exercises that can be used in everyday life to still your mind and overcome worry. Refreshingly free of esoteric or religious overtones.

STILLNESS IS THE WAY
Takes over where MEDITATION A FOUNDATION COURSE leaves off. An advanced course, you are taken through the process of meditation to the simple being of yourself where formal meditation is surpassed. On the way, many of the questions that arise in deeper meditation are answered.

WISDOM AND WHERE TO FIND IT
Takes you far beyond the conventional mind into the realm of pure knowledge.

MAKING LOVE
This book presents a fundamental change in the understanding of love, sex and the nature of man and woman. Its unique insight has transformed the love-lives of thousands around the world.

Only Fear Dies
If vibrant life is truly our birthright why are we ever unhappy? ONLY FEAR DIES examines how we unknowingly cultivate unhappiness in every field of life, and how to be rid of it. A revolutionary perception of human existence.

To Woman In Love
An intimate book of letters between women and a spiritual master whose life has been deeply influenced by his realisation of the divine principle of woman. Touches the heart of the matter for every woman, in and out of love.

To Man In Truth
Profoundly answers the question: How can man's perennial quest for woman and his eternal pursuit of the spirit – historically so at odds – be brought to fulfilment in the modern world?

Raising Children in Love, Justice and Truth
We all love our children. But to bring harmony, love needs to be tempered with justice and knowledge of the truth of life. A practical and enlivening book for parents . . . and everyone.

Available through bookshops or from the addresses below.

Full details and catalogues of Barry Long's books, tapes, videos and teaching programme may be obtained from:

THE BARRY LONG FOUNDATION INTERNATIONAL
Box 5277, Gold Coast MC, Queensland 4217, Australia
BCM Box 876, London WC1N 3XX, England
Suite 251, 6230 Wilshire Boulevard, Los Angeles, CA 90048, USA
or in USA / Canada call 1-800-497-1081
Email: info@barrylong.org
www.barrylong.org